Statutory and Regulatory Property Valuation in Australia

This book explains how property valuation is undertaken for the purposes of taxing land and determining compensation for compulsory acquisition in Australia, known as statutory valuation, together with how property valuation is undertaken for financial reporting purposes or regulatory valuation. Through an analysis of relevant concepts of value, statutes, standards and valuation methods, this book provides the framework within which statutory valuation is undertaken, being an easy guide to follow to the complex and often arcane world of statutory valuation and the globally applicable requirements of regulatory valuation.

For statutory valuation for taxation purposes, the role of mass appraisal valuation approaches for land valuation is considered along with the future role of automated valuation models and the impact of climate change (through increased incidence of bushfires and floods) and systemic shocks such as the COVID-19 pandemic. For statutory valuation for compulsory acquisition compensation purposes, key concepts are considered, with a detailed review of compensation for market value, special value, severance, injurious affection and betterment/worsement. Complex and detailed legislation and case law for the payment of disturbance compensation and indigenous cultural loss compensation are reviewed and applied together with consideration of compensation for non-freehold interests such as leaseholds, easements, strata and substratum.

For regulatory valuation purposes, the interrelated roles of international accounting standards and international valuation standards together with international financial reporting standards are considered, together with an outline of the relevant valuation methods for the assessment of fair value for financial reporting purposes.

An essential textbook for students undertaking property valuation courses at undergraduate and postgraduate levels, this book is also a useful reference for practitioners in banking, accounting, law and other professions dealing with property taxation, compulsory acquisition and financial reporting.

Dr David Parker is an internationally recognised real estate industry expert and highly regarded real estate academic (www.davidparker.com.au), being the author of *Introduction to Property Valuation in Australia* (2024) and editor of *Principles and Practice of Property Valuation in Australia* (2022) and *The Routledge REITs Research Handbook* (2018). He is a regular contributor to property and valuation academic journals, a Visiting Professor at the Henley Business School, University of Reading, a Visiting Fellow at the University of Ulster, an Acting Commissioner of the New South Wales Land and Environment Court and a Sessional Senior Member of the ACT Civil and Administrative Tribunal.

Statutory and Regulatory Property Valuation in Australia

David Parker

Routledge
Taylor & Francis Group
LONDON AND NEW YORK

Designed cover image: Shutterstock

First published 2026
by Routledge
4 Park Square, Milton Park, Abingdon, Oxon OX14 4RN

and by Routledge
605 Third Avenue, New York, NY 10158

Routledge is an imprint of the Taylor & Francis Group, an informa business

British Library Cataloguing-in-Publication Data
A catalogue record for this book is available from the British Library

ISBN: 978-1-032-61692-6 (hbk)
ISBN: 978-1-032-57956-6 (pbk)
ISBN: 978-1-032-61695-7 (ebk)

DOI: 10.1201/9781032616957

Typeset in Times New Roman
by Apex CoVantage, LLC

To Eddie and my twin millennials with excited anticipation as we embark on the next phases of our lives together.

Contents

Foreword

This is a book about economic valuation. The economic value of a thing is measured by the maximum amount of another thing that a person is willing to trade or exchange to obtain, or to give up, the thing to be valued. In a market economy, money is a universally accepted measure of economic value. The amount of money that a person is willing to pay to obtain a thing tells us how much they are willing to give up to obtain the thing. Money acts as a 'numeraire', a tradeable thing in terms of which the relative value of other tradeable things is expressed. When an economist states that, for an individual, a thing has a value of $1,000,000 (being 1,000,000 units of the numeraire of Australian dollars), this means that the individual would be willing to trade or exchange the thing for $1,000,000.

Economic valuation involves defining and measuring the trade-off or exchange between the thing and the units of a numeraire such as money. There are two ways of measuring the trade-off: willingness to pay (WTP) and willingness to accept (WTA). The WTP measure is the maximum amount of money the person would be willing to give up (pay) to obtain a thing. The WTA measure is the minimum amount of money the person would be willing to accept as compensation to forego the thing.

Where the thing is a market commodity, such as private land, in a perfectly competitive private market, the WTP and WTA measures should be close together. Where the thing is a non-market commodity, such as a public good or amenity, if there are private goods that are readily substitutable for the public good, there ought also to be little difference between an individual's WTP and WTA for a change in the public good.

If, however, the thing is a public good that has little or no substitutes, such as an endangered species of plant or animal or a world heritage site, the WTP and WTA measures will differ greatly. The economic value of the thing will therefore depend on which welfare measure is used. Ordinarily, the use of the WTP measure will lead to a lower assessment of value than the use of the WTA measure. The choice of using the WTP measure or the WTA measure will depend on the distribution of property rights. Where a person has a right to the public good or amenity, the correct measure of value is the WTA measure, but where the person has no right, the correct measure of value is the WTP measure.

This book largely concerns the economic valuation of property, including real property (land), property interests such as an easement and a *profit à prendre* in relation to land, and businesses. Property ordinarily is a market commodity. Economic valuation of the property involves measuring the trade-off or exchange between the property and the units of the numeraire of money. The method of measuring this trade-off will vary depending on the property concerned.

For land, which is a market commodity, the market value is the monetary amount for which a willing but not anxious seller would sell the land and a willing but not anxious buyer would buy the land. This is the exchange value – the monetary amount that the landowner is prepared to accept in exchange for the land and the buyer is willing to pay in exchange for the land. This exchange is assumed to occur in an efficient market in which buyers and sellers have access to all currently available information that affects the property.

For other property, other methods of economic valuation may be more appropriate, but the central aim remains the same – to measure the trade-off or exchange between the property concerned and the units of the numeraire of money.

The choice of which valuation method is to be used to measure the trade-off or exchange, and how the chosen method is to be applied to the circumstances, will be influenced, indeed mostly dictated, by the statutory scheme under which the valuation is being undertaken. Valuation for land acquisition is regulated by the applicable statute for the compulsory acquisition of land, such as the *Land Acquisition (Just Terms Compensation) Act 1991* (NSW).

Although not the main focus of the book, the book also considers the economic valuation of non-market commodities. One example is the valuation of native title and indigenous cultural loss. This involves valuing the native title or indigenous cultural holder's loss of welfare. Choices need to be made regarding which of the two welfare measures of WTP and WTA is the more appropriate measure to be used to value the loss of welfare.

The book, therefore, involves a thoughtful exposition of the principles and methods of economic valuation of both market and non-market commodities. The book is divided into two parts: statutory valuation and regulatory valuation. The bulk of the book, Chapters 1–8, address statutory valuation with the final two chapters, Chapters 9 and 10, addressing regulatory valuation.

Statutory valuation involves valuation for compulsory acquisition, rating and taxing purposes. The statutory valuation part of the book is divided into three sections.

The first section, with Chapters 1 and 2, summarises the principles and practice of statutory valuation. One principle considered in Chapter 1, as an example, is the concept of market value. Market value was defined originally by the High Court of Australia in *Spencer v Commonwealth* (1907) 5 CLR 418 but has since been codified in the compulsory acquisition legislation, such as in s 56(1) of the *Land Acquisition (Just Terms Compensation) Act 1991* (NSW).

That definition of market value, however, leaves unarticulated the method by which the market value is to be determined. The most commonly used method is the comparative method of valuation. The market value of the land acquired is inferred from comparable sales of land said to be similar to the land to be valued. If there are no comparable sales, other methods of valuation may be used to determine the market value. These include the capitalisation of income method or discounted cash flow method to derive a market value from the revenue earned from the use of the land; a land residual or hypothetical development method; and a replacement cost method. Such alternative methods of valuation are addressed in Chapter 2 in the discussion of the practice of statutory valuation.

The second section of the statutory valuation part of the book, Chapters 3 to 5, address valuation of the compensation payable for the compulsory acquisition of land. Compensation may be payable for the market value, special value, severance, injurious affection, solatium or disturbance. Different methods of valuation are used to determine the monetary amounts for these different heads of compensation. These different methods are explained.

Chapter 6 discusses compensation for the deprivation of other interests. Compensation may be payable for the deprivation of other interests in land, such as easements, mortgages, mineral rights and profits à prendre, or different land title interests such as strata title and substrata. Compensation may be payable for the extinguishment of a business. And compensation may be payable for native title and indigenous cultural loss. Again, different valuation methods are used for assessing the compensation payable for the deprivation of these interests.

The third section in the statutory valuation part of the book concerns valuation for rating and taxing purposes. Land is used as the basis for rating and taxing of landowners. The basis of value varies depending on the statutes and from state to state. The basis of value might be land value, site value, unimproved value, capital value or some other type of value. The statute is definitive on the applicable basis of value.

Chapter 7 examines the relevant principles for determining the value of land for rating or taxing purposes. Chapter 8 then discusses the practice of valuation for rating and taxing purposes, including the methods of valuation. While there is an overlap with the methods of valuation used for determining the compensation payable for the compulsory acquisition of land, there are some important differences, especially when determining the unimproved value of land that is improved.

The second part of the book concerns regulatory valuation. Regulatory valuation is undertaken for financial reporting purposes rather than for compulsory acquisition, rating or taxing purposes. Chapter 9 explains how regulatory valuation differs from statutory valuation. Regulatory valuation uses a wide range of valuation methods and is not restricted to the methods used for determining compensation for the loss of a property interest. Chapter 9 explores the principles of regulatory valuation, while Chapter 10 discusses the valuation methods used for regulatory valuations such as the income approach and cost approach.

The book thereby serves as an instructive introduction to the two different valuation worlds of statutory valuation and regulatory valuation. As an introductory book, it distils the principles and explains the practice of these two types of valuation. The emphasis is on establishing the framework, rather than detailing the particular statutes and cases on valuation. This framing facilitates a sound understanding of statutory and regulatory valuation, which in turn informs and enables later detailed analysis of particular aspects of statutory and regulatory valuation. This is a strength of the book. The book is a valuable addition to the literature on statutory and regulatory valuation.

The Honourable Justice Brian J Preston AO FRSN SC
Chief Judge of the Land and Environment Court of New South Wales
Sydney
October 2025

Preface

This book is an introduction to two very different valuation worlds in Australia – the complex and often arcane national realm of statutory valuation and the dynamic global empire of regulatory valuation.

With governments around Australia embarking on massive infrastructure building programs, including the Western Sydney Aerotropolis in New South Wales and Victoria's Big Build, the level of acquisition of private property by government is historically high, placing a spotlight on statutory valuation for compulsory acquisition.

Further, taxes on land raise billions of dollars for States, Territories and Councils, highlighting the vital role of statutory valuation in funding government spending programs.

The importance of regulatory valuation is apparent through the periodic financial statements published by every corporate entity in Australia, reporting multi-billion-dollar holdings of land and property based on regulatory valuation.

Whereas only a couple of decades ago the definition of market value provided in *Spencer v Commonwealth* (1907) 5 CLR 418 was omnipresent, it is now only of relevance for statutory valuation, with regulatory valuation focusing on fair value as defined by IFRS and applied through international accounting standards and international valuation standards.

This book focuses on both the principles and practice of statutory valuation and regulatory valuation. Having established the principles of *Spencer* in statutory valuation, the practice of the comparative method of valuation, consistent with the Court's decision in *Adams v Valuer General* [2014] NSWLEC 1005, is considered at length.

The arcane features of statutory valuation, including the role of the public purpose and the before and after method and piecemeal methods of valuation, are explored together with each of the heads of compensation. The tension between value, worth and the concepts of fairness and equity is highlighted with particular regard to compensation for indigenous cultural loss.

The similarly arcane principles of the valuation of land as the basis of taxation are explored, together with the various exemptions, concessions and allowances that have evolved over the last century. The application of mass appraisal, automated valuation models and artificial intelligence is considered in the context of providing a balance for the community between fairness and equity and the cost of providing statutory valuations.

Regulatory valuation is approached from the viewpoint of optimising opportunities for Australia in global trade, with the adoption of IFRS, international accounting standards and international valuation standards ensuring an *apples-with-apples* comparison of land and property valuations in various countries around the world for major international businesses.

The complex interrelationship between IFRS, international accounting standards and international valuation standards as the principles of regulatory valuation is considered, with the application of income and cost approaches to regulatory valuation in practice explored.

The contrast between the complex, often arcane and jurisdictionally specific world of statutory valuation and the globally applicable requirements of regulatory valuation is evident when each is considered side by side. The valuer practising in both statutory and regulatory valuation is effectively operating in two different worlds with two very different sets of principles but some overlaps in practice.

As an introductory textbook on statutory and regulatory valuation, this book is a companion to the Australian valuation textbooks *Introduction to Property Valuation in Australia* and *Principles and Practice of Property Valuation in Australia*, which together provide a comprehensive guide to the principles and practice of property valuation in Australia.

Finally, I express my gratitude to the Honourable Justice Brian Preston for kindly contributing the foreword to this book and to Nick Brunton, Paul Chudleigh, Wayne Lonergan and Tom White for their helpful reviews of various draft chapters.

David Parker
Sydney
October 2025

About the Author

David Parker is an internationally recognised real estate industry expert and highly regarded real estate academic, being a consultant, adviser, mediator and expert witness in property and valuation matters (www.davidparker.com.au).

Dr Parker is Visiting Professor at the Henley Business School, University of Reading and Visiting Fellow at the University of Ulster. He is Acting Commissioner of the Land and Environment Court of New South Wales and Sessional Senior Member of the ACT Civil and Administrative Tribunal, being the former Valuer General of New South Wales and inaugural Professor of Property at the University of South Australia.

With over 40 years' experience in property funds management, real estate investment trusts, valuation standards and statutory valuation, Dr Parker previously held senior executive positions with Schroders Property Fund and ANZ Funds Management.

Holding a BSc, MComm, MPhil, MBA and PhD degrees, Dr Parker is a Fellow of the Royal Institution of Chartered Surveyors, the Australian Institute of Company Directors, the Australian Property Institute and the Australian Institute of Management and a Senior Fellow of the Financial Services Institute of Australasia. He is a member of the Society of Property Researchers, the American Real Estate and Urban Economics Association and the European, American and Pacific Rim Real Estate Societies.

Author of *Introduction to Property Valuation in Australia* (2024) and editor of *Principles and Practice of Property Valuation in Australia* (2022) and *The Routledge REITS Research Handbook* (2018), Dr Parker is also the author of *International Valuation Standards: A Guide to the Valuation of Real Property Assets* (2016) and *Global Real Estate Investment Trusts: People, Process and Management* (2011). Dr Parker has published numerous papers in academic and industry journals and is a regular conference presenter around the world, being an Editorial Board Member for the highly ranked *Journal of Property Research* and *Journal of Property Investment and Finance*.

David Parker may be contacted by email at davidparker@davidparker.com.au.

1 Statutory valuation

Principles

1.1 Introduction

This book is an introduction to two very different valuation worlds in Australia – the complex and often arcane national realm of statutory valuation and the dynamic global empire of regulatory valuation. Statutory valuation is different from regulatory valuation, with the former being based on relevant applicable statutes and being capable of challenge on appeal in Court while the latter is based on relevant applicable regulations and is incapable of challenge in Court (negligence claims excepted).

As an introductory textbook on statutory and regulatory valuation, this book is a companion to the Australian valuation textbooks, *Introduction to Property Valuation in Australia* (Parker, 2024) and *Principles and Practice of Property Valuation in Australia* (Parker, 2022), which provide a wider analysis of key principles underlying property valuation and current techniques and issues in the practice of property valuation for major sectors of the Australian property market.

Statutory valuation in Australia is most commonly encountered in valuations for rating, taxing and acquisition but may arise wherever a statute prescribes a basis of or approach to valuation. Regulatory valuation is most commonly encountered in valuations for financial statements but may arise wherever a regulation prescribes a basis of or approach to valuation.

Rating and taxing of land raise significant amounts of revenue for State Governments and Councils in Australia, based on statutory valuation. For example, the NSW 2025–26 Budget Statement estimates land tax revenue for 2024–2025 to be $8.2 billion with the City of Brisbane Annual Plan and Budget 2024–2025 estimating proposed rates and utility charges to be $1.51 billion.

Acquisition of land, property and businesses commonly arises for major infrastructure projects and may be by agreement or by compulsion, based on statutory valuation. For example, such major infrastructure projects in Australia currently include the multi-billion-dollar Sydney Metro, Melbourne Suburban Rail Loop and Melbourne to Brisbane Inland Rail projects.

Valuation of land and property holdings based on regulatory valuation arises for all reporting entities in Australia that publish financial statements upon which it is reasonable to expect that users will rely for decision making. For example, in June 2025 there are over 2,000 reporting entities listed on the Australian Stock Exchange which each publish financial statements, with the financial statements for Goodman Group, as at 30 June 2024, showing the fair value of total property assets to be $20.2 billion, based on regulatory valuation.

Focusing on the key issues from a valuer's perspective, the concepts and principles of statutory valuation are introduced in Chapters 1 and 2, with statutory valuation for acquisition considered in Chapters 3, 4, 5 and 6 and statutory valuation for rating and taxing considered in Chapters 7 and 8. Regulatory valuation is then considered in Chapters 9 and 10.

DOI: 10.1201/9781032616957-1

This chapter introduces the principles of value in statutory valuation, including the concepts of market value, liberal/conservative estimates of value, highest and best use, potentiality and confirmation of a foresight. This provides the essential foundation principles of statutory valuation that are then applied in the following chapters on statutory valuation.

Chapter 2 considers the practice of statutory valuation, including valuation methods for statutory valuation with an emphasis on the comparative method of valuation, together with statutory valuation conventions such as the before and after method and the piecemeal method. This provides an understanding of the valuation methods and conventions that may be preferred in the following chapters on statutory valuation.

Statutory valuation for land acquisition is considered in Chapters 3, 4, 5 and 6, supported by an evolving example in Appendix 1 of each chapter. Chapter 3 introduces the principles of statutory valuation for land acquisition, including the statutory framework and process, with a focus on the concept of disregard of the scheme in compensation assessment and the distinction between concepts of value and worth. This provides an insight into concepts of fairness and equity in statutory valuation in the context of the requirement to disregard the compulsory nature of the acquisition.

Chapter 4 explores the market value bases of compensation in acquisition for which statutory valuation may be required, including market value of the interest and the related market value compensation concepts of severance and injurious affection. This provides a broad understanding of the heads of compensation for acquisition on a value basis that may arise in statutory valuation.

Chapter 5 considers non-market value forms of compensation including special value, which is noted to rarely arise, payment for solace arising from loss of the primary place of residence and the concept of disturbance for costs and losses suffered as a result of the acquisition. This provides a broad understanding of the heads of compensation for acquisition that are not necessarily on a value basis but which may arise in statutory valuation.

Chapter 6 concludes the chapters considering statutory valuation for acquisition by examining compensation for other interests such as easements and mortgages, mineral rights and profits à prendre, strata title and substrata, businesses and the emerging statutory valuation areas of native title and indigenous cultural loss. This provides an appreciation of the diversity of interests that may be subject to acquisition and so may require valuation for the assessment of compensation.

Statutory valuation for rating and taxing is considered in Chapters 7 and 8. Chapter 7 introduces the principles of land as the basis for taxation through land tax, stamp duty and rates, considering the general principles of the valuation of land for taxation using the New South Wales statutory framework as an example. This provides a general understanding of the often arcane and artificial basis of statutory valuation upon which Government collects very significant amounts of taxation revenue.

Chapter 8 explores the practice of statutory valuation for rating and taxing through exemptions, concessions and allowances that have evolved over the last century, together with contemporary issues such as mass appraisal, automated valuation models and artificial intelligence, the impact of major disasters on valuation for rating and taxing and the litigation process including the valuer as an expert witness. This provides an insight into the continued evolution of statutory valuation to address issues as they arise over time, highlighting the importance of a principles-based approach to statutory valuation.

Regulatory valuation, considered in the context of the preparation of valuations for inclusion in financial reporting, is considered in Chapters 9 and 10. Chapter 9 explores the principles of regulatory valuation with regard to the inter-related roles of international accounting standards, international financial reporting standards and international valuation standards, focusing on the primacy of fair value in International Financial Reporting Standards. This provides an understanding of the fundamental conceptual differences and bases between statutory valuation and regulatory valuation.

Finally, Chapter 10 concludes regulatory valuation by exploring the practice of valuation approaches and methods relevant for regulatory valuation such as the income approach and the cost approach and the various valuation methods there within. This provides an appreciation of the wider range of valuation approaches and methods available to the valuer for regulatory valuation, reflecting that, unlike statutory valuation, there is no requirement for defence in Court (negligence claims excepted).

This book is based on statute, case law and regulations as at March 2025 and those standards and guidance documents published in IVSC (2024) and RICS (2024) and on the API website (accessed January to March 2025). Given their nature, statutes, case law, regulations, standards and guidance documents are dynamic and change continuously. Accordingly, readers should not rely on this book as a current statement of a statute, case law position, regulation, standard or guidance document and should undertake their own enquiries to find the most recent version.

Accordingly, this chapter will now consider the following principles of value in statutory valuation:

- the concept of market value;
- liberal/conservative estimates of value;
- highest and best use and potentiality; and
- confirmation of a foresight.

1.2 Concept of market value

To consider the concept of market value, this section will introduce normative and positive definitions of market value and explore the landmark Australian High Court statutory valuation decision in *Spencer v Commonwealth* (1907) 5 CLR 418 in a contemporary context.

1.2.1 Normative and positive definitions of value

At the core of statutory valuation is the fundamental tension between a positive and a normative definition of value which encapsulates the tension between the taxpayer or dispossessed and the community at large, represented by the Government.

Whipple's (2006) seminal text posits the fundamental distinction between a positive and a normative definition of value as:

- positive concerning what is, being that which has to be accepted as we find it; and
- normative concerning what ought to be, through the prescription of rules.

Further, Whipple (2006), citing Ratcliff in Graaskamp (1977, page 8), states a positive definition of value as follows:

> The most probable price is that selling price which is most likely to emerge from a transaction involving the subject property if it were exposed for sale in the current market for a reasonable time at terms of sale which are currently predominant for properties of the subject type.
>
> (page 106)

Essentially, this is a definition of value in an assumed transaction, being that which is. The seller, effectively, seeks a valuation being what they would be willing to sell the property for or the worth of the property to them, reflecting all the advantages and disadvantages of the property to them. There may be unequal bargaining positions between buyer and seller, such as may arise from compulsion, or the market may not be in equilibrium, but all such features are to be taken into account in the valuation of the property.

For the purposes of taxation or acquisition, the taxpayer or dispossessed would contend that a positive definition of value provides them with a fair and equitable valuation basis for taxation or acquisition. The Government, however, on behalf of the community at large, may contend that a positive definition of value potentially benefits the taxpayer or dispossessed at the cost of the community at large, as they may receive the worth of the property to them.

Conversely, a normative definition of value addresses what value ought to be through a series of 'what ifs' or assumptions. A normative assessment of value may be based on a series of assumptions such as the nature of the buyer and seller, arm's length parties, a given marketing period and so forth. Through the prescription of such rules, an assessment of what the value ought to be may then be determined. In theory, such assumptions, if appropriately set, should be fair to the taxpayer or dispossessed and to the community at large.

For the purposes of taxation or acquisition, the Government, on behalf of the community at large, may contend that a normative definition of value provides the taxpayer or dispossessed with a fair and equitable valuation basis for taxation or acquisition. The taxpayer or dispossessed, however, may contend that a normative definition of value potentially benefits the Government or community at the cost of the taxpayer or dispossessed, who is deprived of the worth of the property to them.

Herein lies the fundamental tension between a positive definition of value, that the taxpayer or dispossessed may see as fair and equitable and providing an expression of worth to them, and a normative definition of value, that the Government may see as fair and equitable and not enriching the taxpayer or dispossessed at the expense of the community at large.

Significantly, statutory valuation in Australia is based on a normative definition of value arising from the Australian High Court decision in *Spencer v Commonwealth* (1907) 5 CLR 418. Accordingly, therefore, in the absence of legislative amendment to move from a normative definition of value to a positive definition, there will always be taxpayers or dispossessed who are unhappy and consider the taxation and acquisition system is both unfair and inequitable.

1.2.2 **Spencer v Commonwealth** ***(1907) 5 CLR 418***

The landmark decision of the High Court of Australia in *Spencer v Commonwealth* (1907) 5 CLR 418 (*Spencer*) provides the fundamental basis of valuation for statutory valuation in Australia. Following the cessation of colonial status with the creation of the Commonwealth of Australia upon Federation on 1 January 1901 and separation from Great Britain, the High Court of Australia was only 4 years old when it heard an appeal on a compulsory acquisition matter.

The appeal concerned the compulsory acquisition by the Commonwealth of 6 acres, 1 rood and 2 perches of virtually vacant land in North Fremantle for the construction of a fort for the defence of Fremantle harbour, motivated by the Russo-Japanese War of 1904–5, being described as follows:

> The land consists of sand-hummocks overlooking the Indian Ocean. It has no grass; and it is useless in its present condition for any purpose of production.

with the valuation date set as 1 January 1905.

The Commonwealth offered Spencer £2,641 in compensation and Spencer sought compensation of £10,000. Spencer brought an action in the High Court claiming compensation of £10,000 and the Commonwealth paid £3,000 into the Court without denying liability. The originating action was heard by Higgins J, with the appeal against his decision heard by Griffith CJ, Barton J and Isaacs J, who awarded Spencer the £3,000 already paid into Court.

The bench was particularly notable, comprising Griffith CJ, who was generally claimed to be the principal author of the Constitution of Australia and the first Chief Justice of Australia, Barton J, who had previously been the first Prime Minister of Australia from 1901 to 1903, and Isaacs J, who became the first Australian-born Governor-General of Australia in 1930.

It is important to note that the High Court was considering value in the context of compulsory acquisition. Spencer and the Commonwealth would presumably have had a period of negotiation to try and agree a value for the land. Such a negotiation would have presumably allowed Spencer to contend that which he considered to be the value of the land to him in order to entice him to sell and the Commonwealth would contend that which they thought should be paid for the land. As a compulsory acquisition process, there was no opportunity for the property to be marketed and for the parties to see what, if any, bids may be made for the land as would occur in a normal sale process. Further, the negotiations could not continue indefinitely as the fort project needed to proceed. Accordingly, as the two positions did not meet, the negotiation presumably would have been unsuccessful and the acquisition became compulsory.

As the parties were unable to agree, it ultimately came to the High Court to determine the matter. By this point, title would presumably have transferred from Spencer to the Commonwealth and all that was required to be decided was the value of the land that was to be paid as compensation. Significantly, the High Court did not contemplate the value of the land to Spencer nor the value of the land to the Commonwealth, nor did it contemplate that value at which Spencer and the Commonwealth may have been able to agree (which would have been positive definitions of value), but instead formulated its own approach to the determination of value (which was a normative definition of value) attempting to strive for a balance that was fair to both parties – to Spencer as the dispossessed owner and to the Commonwealth, which represents its taxpayer citizens, being the community at large.

The significance of the decision lies in the way in which the Judges constructed the concept of value:

> In my judgement, the test of value of land is to be determined, not by inquiring what price a man desiring to sell could actually have obtained for it on a given day, ie, whether there was in fact on that day a willing buyer, but by inquiring "What would a man desiring to buy the land have had to pay for it on that day to a vendor willing to sell it for a fair price but not desirous to sell?" It is, no doubt, very difficult to answer such a question, and any answer must be to some extent conjectural. The necessary mental process is to put yourself as far as possible in the position of persons conversant with the subject at the relevant time, and from that point of view to ascertain what, according to then current opinion of land values, a purchaser would have had to offer for the land to induce such a willing vendor to sell it, or, in other words, to inquire at what point a desirous purchaser and a not unwilling vendor would come together.
>
> (Griffiths CJ)

> And I should say, in view of the many authorities cited and upon the sense of the matter, that a claimant is entitled to have for his land what it is worth to a man of ordinary

> prudence and foresight, not holding his land for merely speculative purposes, nor, on the other hand, anxious to sell for any compelling or private reason, but willing to sell as a business man would be to another such person, both of them alike uninfluenced by any consideration of sentiment or need.
>
> (Barton J)

> To arrive at the value of the land on that date, we have, as I conceive, to suppose it sold then, not by means of a forced sale, but by voluntary bargaining between the plaintiff and a purchaser, willing to trade, but neither of them so anxious to do so that he would overlook any ordinary business consideration. We must further suppose both to be perfectly acquainted with the land, and cognizant of all circumstances which might affect its value, either advantageously or prejudicially, including its situation, character, quality, proximity to conveniences or inconveniences, its surrounding features, the then present demand for land, and the likelihood, as then appearing to persons best capable of forming an opinion, of a rise or fall for what reason soever in the amount which one would otherwise be willing to fix as the value of the property.
>
> (Isaacs J)

together with:

> In order that any article may have an exchange value, there must be presupposed a person willing to give the article in exchange for money and another willing to give money in exchange for the article.
>
> (Griffiths CJ)

> . . . value implies the existence of a willing buyer as well as of a willing seller . . .
>
> (Griffiths CJ)

> Prosperity unexpected, or depression, which no man would ever have anticipated, if happening after the date named, must be alike disregarded.
>
> (Isaacs J)

> . . . the all important fact on that day is the opinion regarding the fair price of the land, which a hypothetical prudent purchaser would entertain, if he desired to purchase it for the most advantageous purpose for which it was adapted. The plaintiff is to be compensated; therefore he is to receive the money equivalent to the loss he has sustained by deprivation of his land, and that . . . cannot exceed what such a prudent purchaser would be prepared to give him.
>
> (Isaacs J)

Within the Judges' construction of the concept of value, the following elements may be identified:

- **an estimated amount:** *could actually have obtained for it, have had to pay* (Griffith CJ), *value of the land, the fair price of the land* (Isaacs J);
- **an exchange:** *a man desiring to buy the land have had to pay for it . . . to a vendor willing to sell, may have an exchange value* (Griffith CJ), *willing to trade* (Isaacs J);
- **a valuation date:** *on a given day, on that day* (Griffith CJ), *on that date, prosperity unexpected, or depression,. . . if happening after the date named* (Isaacs J);

- **a willing buyer:** *a man desiring to buy the land, a desirous purchaser, willing buyer* (Griffith CJ), *voluntary bargaining, willing to trade, hypothetical prudent purchaser* (Isaacs J);
- **a willing seller:** *a vendor willing to sell it for a fair price, a not unwilling vendor* (Griffith CJ), *voluntary bargaining, willing to trade* (Isaacs J);
- **a knowledgeable and prudent buyer and seller:** *to a man of ordinary prudence and foresight, willing to sell as a business man would be to another such person* (Barton J), *perfectly acquainted with the land, hypothetical prudent purchaser* (Isaacs J);
- **an absence of compulsion:** *but not desirous to sell* (Griffith CJ), *nor, on the other hand, anxious to sell for any compelling or private reason, uninfluenced by any consideration of sentiment or need* (Barton J), *not by means of a forced sale, overlook any ordinary business consideration* (Isaacs J); and
- **an assumption of highest and best use:** *the most advantageous purpose for which it was adapted* (Isaacs J).

Accordingly, at the beginning of the last century, within the judgements in *Spencer*, the key elements of the concept of that which is now known as market value may be identified, in the form of a normative definition.

1.2.3 **Spencer** *in a contemporary context*

It should, however, be noted that the judgements in *Spencer* were in the context of a vacant block of land on the coast of Western Australia in 1905 such that, while as a concept it is still of relevance today, care is required in the interpretation of language for application in this century:

- **an estimated amount:** the use of the term estimate reinforces that a valuation is a matter of opinion rather than a matter of fact, representing the point of intersection between the opinions of the willing buyer and the willing seller. It may be contended that the estimated amount is the result of a 'plain vanilla' or 'normal' transaction without any special terms, financing, leaseback, concessions and so on. It may further be contended that such a 'normal' transaction is between unrelated parties, being an arm's length transaction rather than a related party transaction. However, a 'normal' transaction may have differing characteristics for differing types of property in differing sub-markets. Accordingly, the estimated amount may be contended to be upon whatever characteristics (such as financing or concessions) as are typical for that sub-market;
- **an exchange:** the existence of an exchange is fundamental and an assumed exchange has to happen. It is, therefore, not permissible for an assumed willing seller in a declining or depressed market to contend that they would simply not have sold the property. While not necessarily an issue in statutory valuation for taxation, this requirement goes to the issue of fairness and equity in dispossession where the dispossessed party may be dispossessed at the bottom of the market while, in reality, they would not have sold and waited until the market improved.

  Further, consistent with the existence of an exchange, is the assumption of a clearing price. Even in the most depressed market, there will be a price at which a property would trade between assumed parties if not between actual parties;
- **a valuation date:** the assessment of value is temporal, being reflective of market conditions applicable at the valuation date. Accordingly, care is required when markets are moving rapidly upwards or downwards, especially when at an inflection point or when a specific event occurs such as a rezoning or an interest rate change;

- **a willing buyer/seller:** consistent with the knowledgeable and prudent condition, a rational buyer/seller may be contended to be motivated but not compelled to buy/sell which is consistent with capital market theory and finance theory. However, it is inconsistent with behavioural finance theory and the tendency for a willing buyer/seller to be influenced by anchoring, emotional involvement and so on, therefore disregarded. Effectively, the willing buyer/seller may be assumed to be rational and emotionally detached.

  Significantly, while the willing buyer/seller are hypothetical, each requires to be characterised. While only one willing buyer/seller is required, a single party should not be contemplated as this may lead to the reflection of idiosyncratic behaviours in the value estimated. The willing buyer may be considered as a group (such as institutional investors for a high rise CBD office tower or private investors for a single shop with residence above) and the willing seller may be considered as that group to which the taxpayer or dispossessed belongs. Consideration as a group permits the valuer to avoid focusing on idiosyncratic influences on value and only consider systemic influences for the identified group;
- **a knowledgeable and prudent buyer and seller:** this condition may be considered to relate to awareness of information (knowledge) and use of information (prudence). As such, it should be interpreted relative to a vacant coastal block in 1905 where the amount of information available was limited (situation, character, quality, proximity to conveniences and inconveniences, surrounding features, then present demand per Issacs J) and the awareness and use of such information to become knowledgeable was not challenging.

  In the context of properties today, that level of information awareness (preceded by information collection and analysis) and information use may be contended to vary significantly by property type and sub-sector. While a modest amount of information awareness and use may be anticipated for a suburban housing block (area, title, zoning, development constraints, etc.), a significantly higher level of information awareness and use may be anticipated for a complex major investment property (being more in the nature of a due diligence exercise) in order for the willing buyer to become knowledgeable. However, the knowledge of all matters is not assumed to be perfect (Biscoe J in *McDonald v Roads and Traffic Authority of NSW* [2009] NSWLEC 105 at 35).

  While the market may be assumed to be efficient, in that information is impounded into price, it remains a property market with the limitations on efficiency typical of property markets. As with other finance theories, care should be taken in attempting to apply concepts like the efficient market hypothesis, with a greater focus on the likely level of information awareness and use by the identified buyer group acting prudently to be preferred.

  It should also be noted that prudence has a temporal dimension, with that which may be considered prudent at the peak of a market differing from that which may be considered prudent when the market has collapsed. The assessment of prudence by the willing parties is, therefore, a snapshot rather than a video;
- **an absence of compulsion:** this condition is a keystone in statutory valuation in Australia, requiring the lack of choice to sell and the impact thereof on the actual seller to be disregarded. At the time of the High Court decision, this condition differed from other jurisdictions in the British Empire, with Baum et al (2014) noting that in Great Britain:

  > Prior to 1919 it had become accepted practice to add 10% to the agreed quantum of compensation as a consolation for compulsory expropriation.
  >
  > (page 226)

The disregard of compulsion is a significant element in the normative definition of value developed in *Spencer*, effectively potentially disadvantaging the dispossessed for the benefit of the community at large. Over a century later, a common complaint of some dispossessed remains that they would not have sold the property but for the acquisition, often compounded by the claim that they definitely would not have sold it for the amount received, which contributes to ongoing tension in the acquisition system; and

- **an assumption of highest and best use:** the reference, by Isaacs J, to *the most advantageous purpose for which it was adapted* is also a fundamental concept in the normative definition of value for statutory valuation purposes, presaging that which is known today as highest and best use. While Isaacs J does not offer any criteria upon which the *most advantageous purpose* may be determined, subsequent case law provides clarification and is considered in Section 1.4.1.

With the exception of a reference to highest and best use, an arm's length transaction and the assumption of pre-sale marketing, the normative definition of value from *Spencer*, for the purposes of statutory valuation in Australia, is summarised by Rost and Collins (1990) in their seminal work as follows:

> . . . the High Court defined value, in effect, as the price which a property could be expected to realise if sold by a willing but not anxious seller to a willing but not anxious buyer at the date at which the value is required to be ascertained. Thus the land valuer is obliged to visualise the ideal sale, where both parties to the transaction are prudent, well informed, and unaffected by any abnormal influence that would deprive the sale of weight as a criterion of value.
>
> (page 37)

1.3 Principle of liberal/conservative estimate

The concept of the liberal/conservative estimate in statutory valuation requires that valuations should favour the taxpayer (i.e. be a conservative estimate) in taxation/revenue matters and favour the dispossessed (i.e. be a liberal estimate) in acquisition matters. Favour should be construed in the sense of rounding up/down or favouring the upper/lower end of a range rather than simply selecting the highest/lowest valuation.

The concept is often referred to as the *Executor Trustee principle*, being founded in Dixon's dicta in *Commissioner of Succession Duties (SA) v Executor Trustee and Agency Co of South Australia* (1947) 74 CLR 358:

> I should like, however, to add for myself that there is some difference of purpose in valuing property for revenue cases and in compensation cases. In the second the purpose is to ensure that the person to be compensated is given a full money equivalent of his loss, whilst in the first it is to ascertain what money value is contained in the asset so as to afford a proper measure of liability to tax. While this difference cannot change the test of value, it is not without effect upon a court's attitude in the application of the test. In a case of compensation, doubts are resolved in favour of a more liberal estimate: in a revenue case, of a more conservative estimate.
>
> (373–374)

and was summarised in *Rigby v Secretary to the Department of Sustainability and Environment* [2012] VSC 427 as follows:

> Dixon J identified a different purpose in valuing property in revenue cases from compensation cases: in the compensation cases the purpose was to ensure that the person to be compensated was given a full equivalent of his or her loss, while in the revenue cases it was to ascertain what money value was plainly contained in the asset so as to afford a proper measure of liability to tax.
>
> (28)

Rost and Collins (1990) poetically note:

> It has become axiomatic that in compensation cases the Courts will not endorse any niggardly approach in their adjudication upon claims. Where doubts exist the tendency is in the direction of liberality, rather than conservatism. The valuer must avoid stinginess, but this does not mean that liberalism is to be extended beyond the limits of reality.
>
> (page 551)

Accordingly, in taxation/revenue matters the valuer should err on the side of pessimism and in dispossession matters err on the side of optimism.

1.4 Principles of highest and best use and potentiality

Often referred to as the *Turner* principle (after *Turner v Minister of Public Instruction* (1956) 95 CLR 245), for statutory valuation purposes a property is to be valued with all its potential. This may include short-term potential and long-term potential with short-term potential considered below as highest and best use and long-term potential as potentiality.

1.4.1 Principle of highest and best use

That which is today known as the concept of highest and best use was identified as an element in the concept of market value in the judgements in *Spencer*:

> the most advantageous purpose for which it was adapted
>
> (Isaacs J)

with each party taking into account:

> not only the present purpose to which the land is applied, but also any more beneficial purpose to which, in the course of events at no remote period it may be applied, just as an owner might do if he were bargaining with a purchaser in the market.
>
> (Isaacs J)

As Rost and Collins (1990) note:

> The valuation of land for its highest and best use demands that due weight be given to its potential utility and to the probability of consent being given for such potential use.
>
> Potential utility refers to a latent and more beneficial use which may be capable of coming into being, or a use existing as a possibility only.
>
> (page 91)

Highest and best use is a use of the land, in the short-term future, which may be likely to be achievable. A common example is land currently used for grazing but zoned for low-density residential use, for which the highest and best use would be low-density residential development and for which the value would be determined on the basis of land used for low-density residential development.

The highest and best use does not need to be capable of specific definition (*precise highest and best use*) but *might be identified in general terms, such as an intense use for office, residential, hotel, club or retail purposes* (*ISPT Pty Ltd v Melbourne City Council* [2008] VSCA 180).

While there may not be a statutory definition of highest and best use, various cases in Federal, State and Territory Courts have considered aspects of that which constitutes highest and best use.

Sheahan J considered the principles of highest and best use extensively in *Limina Holdings Pty Ltd ITF Galileo Superannuation Fund v Valuer General of New South Wales* [2019] NSWLEC 110, noting:

> 91. There is no statutory definition of "highest and best use", but the Australian Property Institute considers it to be "the use of the property that maximizes its potential which is physically possible, legally permissible, and financially feasible" (Exhibit A2, page 19, sec 8.5).
>
> 110. 63 As Jacobs J pointed out in [Adelaide Clinic], at 415, highest and best use is not the value of land assuming there are no planning or like controls. Rather it is the most profitable use having regard to the physical, economic and legal constraints on the use of the land.
>
> 113. I infer from the writings of judges as distinguished as Dixon CJ, Isaacs J, Wells J, Else-Mitchell J, Hill J, and Biscoe J, the importance of recognising, for the purposes of identifying the highest and best use of relevant land, underpinning concepts of feasibility, realism, practicality, and reasonableness, reviewed in the context of all the "circumstances" of the land – what Jacobs J called the "planning and all other relevant factors affecting its present and future potential".

Further, the High Court echoed the Australian Property Institute's definition of highest and best use in *Valuer-General Victoria v WSTI Properties 490 SKR Pty Ltd* [2025] HCA 23:

> In modern terminology, "adapted" or "highest and best use" of land means the most valuable use of the land, objectively ascertained by reference to the hypothetical buyer, which is "physically possible, legally permissible and financially feasible".
>
> (37)

Therefore, physically possible, legally permissible and financially feasible comprise those constraints to the realisation of highest and best use. Continuing the example of land currently used for grazing but zoned for low-density residential use, realisation of the value for highest and best use for low-density residential development is dependent upon such use being:

- physically possible – the land itself must be capable of residential development, such as in terms of slope, size, shape and so forth;
- legally permissible – the land is zoned for residential development but appropriate approval would be required, raising the question of whether or not such approval would be likely to be granted; and

- financially feasible – requiring a feasibility study to determine if the land could be profitably developed for residential use – if not, the value of the land is likely to be closer to that of grazing land.

It may be contended that physically possible requires focus on that use which would be considered reasonable by market participants, requiring consideration of such physical issues as being capable of real-world creation given such constraints as the location, size and topography of the land.

Similarly, consideration of legally permissible may be contended to focus on planning aspects, such as use, nature of development, density, height and so forth, but may also include permissibility within the form of title held, which is particularly significant for leasehold interests, licences and other forms of title that are not freehold title.

Further, consideration of financially feasible may be contended to focus on acceptability of return as a relative measure, being relative to the existing use, rather than as an absolute measure, consistent with the assumption of the rational hypothetical buyer.

It should be noted that all three constraints require capability of achievement, rather than fulfilment, if the value for highest and best use is to be realised, such that a proposed use which is physically possible and legally permissible may not be the highest and best use if it is not also financially feasible. Effectively, case law suggests that the assumption of highest and best use should reflect a real-world scenario that could actually occur rather than a hypothetical scenario that would be unlikely to or would not occur in reality.

While the High Court's approach to the determination of highest and best use in *Valuer-General Victoria v WSTI Properties 490 SKR Pty Ltd* [2025] HCA 23 is closest to that definition adopted in non-statutory valuation, other respected authors have proffered further wider criteria.

Jacobs (2015) notes that, in *Department of Transportation and Development v Hammons* 550 So 2d 767 (La App 2 Cir) (1989), the Court identified the following elements that contribute to the determination of highest and best use:

- market demand;
- economic development in the area;
- specific plans to develop the land for that use;
- uses permitted by zoning ordinances;
- proximity to areas already developed in a manner compatible with the intended use; and
- the use to which the property is being put at the time of the taking (page 629).

Brown (2009) further notes that the following elements are required to establish highest and best use:

- the best use must be legal – it must come within the planning and building regulations;
- the best use must be within the realm of probability – it must be likely;
- the best use must not be speculative, conjectural, unrealistic or improbable;
- the best use must be of a kind to come within the imagination of a hypothetical purchaser who appreciates and understands all the problems involved in a change of use but is nonetheless confident that the problems can be overcome or solved; and
- the best use can be carried out within a reasonably short time (page 133).

In terms of valuation application, use of relevant comparable sales for other land exhibiting the highest and best use would be preferable. In the absence of comparable sales of land for the highest and best use, adjustment of other relevant sales would be required. Continuing the example of land currently used for grazing but zoned for low-density residential use, in the absence of comparable sales of low-density residential land, the valuer has two potential adjustment options:

- a bottom up premium – using sales of grazing land and adjusting upwards for the potential for low-density residential development, with all the risks inherent in achieving development approval and a profitable development being included within the premium adopted; or
- a top down discount – using sales of medium-density residential development land and adjusting downwards for the potential for low-density residential development, with all the risks inherent in achieving development approval and a profitable development being included within the discount adopted.

It is important to view the potential achievement of the constraints to highest and best use through the eyes of the hypothetical purchaser in the actual market as at the valuation date, the question being not whether the highest and best use will be achieved but whether the hypothetical purchaser would view it as being achievable in the context of all other information available.

However, for compulsory acquisition compensation, it should be noted that compensation on the basis of highest and best use rather than current use will generally preclude a claim for disturbance compensation to the extent that it is inconsistent with a claim for market value based on the potential use of the land, reflecting that, in realising highest and best use value, current use would be discarded.

1.4.2 Principle of potentiality

The concept of potentiality, being that a property should be valued at the relevant date in its existing condition with all its 'potentialities', was established in *Yates Property Corporation Pty Ltd (in liq) v Darling Harbour Authority* (1991) 24 NSWLR 156 at 175–176.

The fundamental difference between the concept of highest and best use and the concept of potentiality is temporal, with short-term potential considered above as highest and best use and long-term potential considered below as potentiality.

Highest and best use as short-term potential was referred to in *Spencer*, with each party taking into account:

> not only the present purpose to which the land is applied, but also any more beneficial purpose to which, in the course of events at no remote period it may be applied, just as an owner might do if he were bargaining with a purchaser in the market.
>
> (Isaacs J)

Where the period may be more remote or long term, the Court stated in *Turner v Minister of Public Instruction* (1956) 95 CLR 245:

> Future advantages or potentialities must not be excluded.
>
> (268)

and in *Beard v Director of Housing (Tas)* [1961] Tas SR 141:

> If the land to be valued enjoys in whole or in part a special potentiality which is not reflected in its present use, then it is clear that in the notional price which the court must find it should

> include full allowance for that potentiality, but such allowance will not exceed that added value which a purchaser of the land as a whole would presently pay for the land in contemplation of its advantages: Turner v Minister of Public Instruction (1956) 96 CLR 245.

The temporal distinction between highest and best use and potentiality was also evident in *McCann v Roads Corporation* [2011] VSC 96:

> The potential highest and best use of land may itself be to hold the land for a potential future use. Thus, there is a distinction which may be drawn between land ripe for subdivision and land having the potential for future subdivision.
>
> (25)

and in *Minister v Matford Nominees Pty Ltd* [1973] 2 NSWLR 58, Else-Mitchell J stated:

> the value of land at any point in time will reflect its potentiality for the best or highest or most profitable use to which it can be put. This potentiality will arise before actual development or redevelopment is undertaken and, in the case of en globo lands or broad acres, must not be regarded as non-existent until approval for development or redevelopment has in fact been granted.
>
> (59–60)

A common physical manifestation of potentiality is in the practice of land banking by developers on the outer fringe of towns and cities. As urban development spreads outwards, developers may speculatively acquire outer fringe rural land on the basis that such land may, at some unknown date in the future, be rezoned for some form of residential development. Such land may be unlikely to fulfil the constraints of physically possible, legally permissible and financially feasible to establish a highest and best use as residential development land but the land exhibits potentiality, being the potential for such highest and best use at some point in the future. It may be contended, therefore, that a hypothetical purchaser may be likely to pay a premium over the value of land for its current zoning in recognition of such potentiality.

The challenges of reflecting the potentialities arising from the rezoning process were considered by the NSW Land and Environment Court in four compulsory acquisition cases concerning the North West Rail Link in 2014:

Attard v Transport for NSW [2014] NSWLEC 44 (Biscoe J)
Chircop v Transport for NSW [2014] NSWLEC 63 (Biscoe J)
Bonomo v Transport for NSW [2014] NSWLEC 25 (Sheahan J)
De Battista v Transport for NSW [2014] NSWLEC 39 (Pain J)

A significant issue in such consideration was the extent to which the land was 'ripe' for the realisation of the potentiality, being the relativity between certainty and uncertainty of realisation, with Biscoe J noting in *Attard v Transport for NSW* [2014] NSWLEC 44:

> 61 I have earlier concluded that at the acquisition date it was virtually certain Riverstone East precinct in which they are located would be rezoned for residential subdivision under the Growth Centres SEPP, the likely rezoning would be R2 and occur in late 2014/early 2015, and the subject land would then be ripe for residential subdivision redevelopment which would likely occur 15–18 months after rezoning.

As Valuer General (2021) notes:

> The cases concerned the North Western Growth Centre and its association with the North Western Rail Link, a 23-kilometre train line between Bella Vista and Epping which necessitated the compulsory acquisition of 60 homes and 30 businesses. Predominantly rural and/or rural-residential land, which would ultimately, at some undefined point in the future, have become higher order use residential development land but for the acquisition, was compulsorily acquired at various stages of the rezoning journey with the Court required to determine market value. The Court grappled with the challenge of recognising potentiality in market value during the rezoning process, when relevant sales may recognise a range of issues including potentiality.

The judgements examined the comparable sales relied upon by the acquiring authority and the dispossessed in each case and detailed information on the extent to which potentiality was being realised, including Government announcements on the provision of infrastructure, particularly sewage and water, together with approval of the North West Rail Link, Premier and Minister's statements, State Budget, Government Fact Sheets, road upgrade announcements and Community Announcements.

Having regard to the judgements and, in the context of the Western Sydney Aerotropolis, a major infrastructure project which has been in the process of evolution for around 50 years leading to announcement in April 2014, the NSW Valuer General undertook a detailed study, *Review of the impact of rezoning potentiality on land values* (Valuer General, 2021).

Valuer General (2021) found that the hypothesis that the land value impact of the rezoning process follows four key stages to be generally supported, with the four key stages illustrated in Figure 1.1.

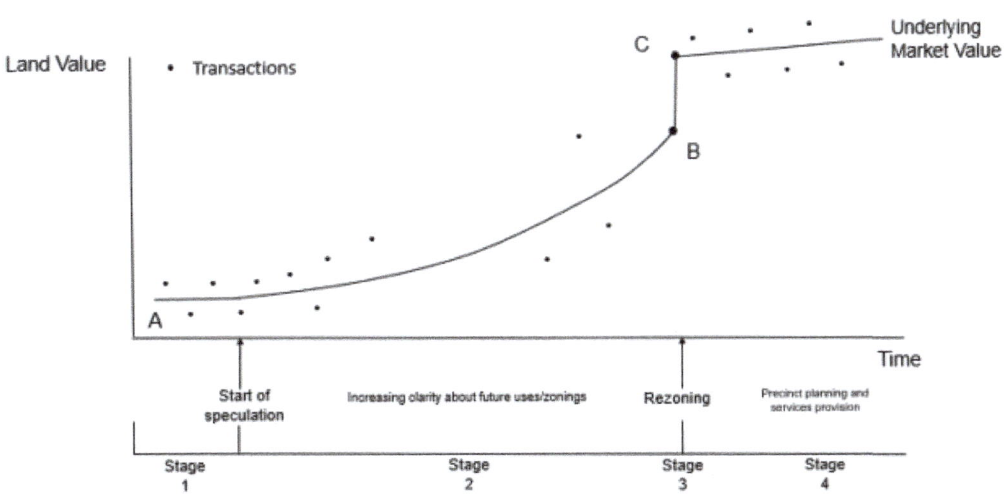

Figure 1.1 Hypothesis of the impact on land value of the potentiality for rezoning.

Source: Valuer General (2021)

Valuer General (2021) outlines the four stages as follows:

- Stage 1 – being the normal market period prior to any rezoning potentiality;
- Stage 2 – being the period from when initial speculation about rezoning starts up until the rezoning occurs;
- Stage 3 – being the point at which the rezoning is formalised; and
- Stage 4 – being the period where precinct planning and services provision occurs.

1.4.2.1 Stage 1

Extending for the period from a commencing point, A, where the market is stable and in equilibrium, to the start of speculation that rezoning may occur.

For example, in this period, land may be zoned rural and values for rural land may increase slightly over time, with transactions occurring on a regular basis and generally evenly distributed around underlying market value.

In Stage 1, transactions and underlying market value are generally aligned such that evidence to support the assessment of land value is readily available and generally consistent.

1.4.2.2 Stage 2

Commences when speculation about rezoning starts, with vendors and purchasers incorporating potentiality within the price formation process. Stage 2 ends at point B when the rezoning is formalised.

For example, in this period, there may be speculation that the land zoned rural may be rezoned for residential purposes. Market participants do not know when this rezoning may occur or exactly what type of residential use may be permitted – however, they anticipate rezoning is going to happen at some point in the future and that their land will then have a higher order use and so be more valuable than use for rural purposes.

This may result in transactions occurring on a less frequent basis, as vendors hold their land in anticipation of a value rise and specific circumstances may influence prices such as vendors keen to sell for personal reasons, purchasers keen to buy to amalgamate sites and so forth.

Given the limited and inconsistent information in the market, vendors and purchasers may interpret this differently, leading to a wide distribution of prices which generally indicate an upward trajectory in land value.

While the transactions may be widely distributed and challenging to analyse, it may be anticipated that the underlying land value will start to increase at a greater rate as clarity about the rezoning increases. This may follow press speculation, Ministerial announcements, Council meetings, rumours and so forth. Depending on the nature of the information, the value increase may gain greater momentum and become exponential.

In Stage 2, transactions and underlying market value are generally not aligned such that evidence to support the assessment of land value is not readily available and not generally consistent.

1.4.2.3 Stage 3

Starts at point B, where the rezoning is formalised and results in the realisation of potentiality being manifest in transactions at point C.

Up to point B, the rezoning was anticipated but not certain. At point C, the rezoning is certain and, therefore, the risk of the rezoning not occurring is removed and land value jumps accordingly.

In Stage 3 at point C, transactions and underlying market value are generally aligned such that evidence to support the assessment of land value is readily available and generally consistent.

1.4.2.4 Stage 4

Starts at point C, where the rezoning is certain and continues into the future.

For example, in this period, land will have been rezoned residential and values will have risen to reflect the certainty of use as residential. Following rezoning, there may be a period where precinct planning and services provision occurs, with the exact scheduling of same uncertain.

This Stage may be a hybrid of Stage 1 and Stage 2, with transactions reflecting the certainty of zoning but with the potentiality of precinct planning and services provision.

Accordingly, land value may increase slightly over time, with transactions occurring on a regular basis, but the uncertainty may result in a mildly uneven distribution around underlying market value.

In Stage 4, transactions and underlying market value are generally aligned such that evidence to support the assessment of land value is readily available and generally consistent.

1.4.2.5 Summary – potentiality

Valuer General (2021) concludes that the hypothesis that the land value impact of the rezoning process follows four key stages is found to be generally supported from transactions analysed. Further, it is in Stage 2 that the greatest challenge may arise in determining land value for the purposes of statutory valuation. This may be particularly important for undertaking statutory valuations for dispossession, where rezoning may be an increasingly closer potentiality but not yet an actuality.

Given the long time periods over which major infrastructure projects may be known in the community to be likely, great care is required in the analysis of comparable sales (which may be very disparate) to determine how much potentiality is exhibited in such a sale. As Figure 1.1 shows, values trend upwards towards point B (the announcement of the rezoning) indicating an increasing willingness by the market to pay for potentiality. Accordingly, regard should be given by the valuer to the impact of potentiality where the local situation suggests it may have an impact on value.

As with highest and best use, in the absence of comparable sales evidencing the impact of potentiality, value may be determined by addition of a bottom up premium to current zoning land value or deduction of a top down discount from higher order zoning land value, with the level of premium/discount reflecting the evolution in the relevant planning instruments, other necessary changes and the passage of time before such potentiality may be realised.

1.5 Principle of confirming a foresight

While the general principle is to have regard only to matters before and as at the date of valuation, regard may be given to matters that are foreseen at the date of valuation and reflected in the valuation and which are then later confirmed by events.

Such matters may include regard to the sale of comparable blocks of land for which terms had been agreed but which had not settled as at the date of valuation, regard to a rising/falling interest rate environment or regard to changes in zoning or utility provision that were imminent and widely known but not yet actual.

Foresight may be significant in markets which may be rising or falling with little if any comparable sales evidence as at the date of valuation. The valuer may posit that the market had risen or fallen at the date of valuation and use comparable sales evidence from after the date of valuation to confirm that their foresight was correct.

Such matters of foresight are those matters to which the hypothetical willing but not anxious purchaser may have had regard when agreeing to a transaction price with the hypothetical vendor.

In *Housing Commission of NSW v Falconer* [1981] 1 NSWLR 547, Hope JA stated:

> However there are many decisions, including decisions of the High Court, in which it has been held that evidence of future events is admissible not to provide a hindsight, but to confirm a foresight . . .
>
> The extent to which foresight coincided with or approached fact must be a question of fact – assuming there is evidence to support a conclusion . . .
>
> (558–559)

In *Roads Corporation v Love* [2010] VSC 32, Osborne J summarised the principle of confirming a foresight as follows:

> Regard may be had to evidence of events subsequent to the acquisition as confirming or tending to confirm the foresight which the hypothetical purchaser would have had at the relevant date.
>
> (121)

To effectively rely on the foresight principle, the valuer should clearly state that which is being foreseen and relied upon as at the date of valuation in their valuation report such that it can be clearly identified for confirmation by later events. Simply adopting a retrospective inference after the valuation date that such events were foreseen in the valuation is unlikely to be sufficient to rely upon, particularly if the foreseeability period is long.

1.6 Summary and conclusions

Focusing on the key issues from a valuer's perspective, this chapter and Chapter 2 introduce the concepts and principles of statutory valuation, with statutory valuation for acquisition considered in Chapters 3, 4, 5 and 6 and statutory valuation for rating and taxing considered in Chapters 7 and 8. Regulatory valuation is then considered in Chapters 9 and 10.

This chapter sought to introduce the principles of statutory valuation, including the concepts of market value, liberal/conservative estimates of value, highest and best use, potentiality and confirmation of a foresight. Market value, based on the seminal case of *Spencer v Commonwealth* (1907) 5 CLR 418, is a foundation concept in statutory valuation in Australia, creating the fundamental tenets of the willing but not anxious knowledgeable and prudent hypothetical vendor and purchaser, an exchange, a valuation date, an absence of compulsion and highest and best use. Significantly, the *Spencer* definition of value is explored as a normative definition of value and not a positive definition of value, creating tensions in fairness and equity between the taxpayer or dispossessed and the government or acquiring authority representing the community at large.

The concept of a conservative estimate favouring the taxpayer in taxation/revenue matters and a liberal estimate favouring the dispossessed in acquisition matters, being the *Executor Trustee principle*, was explored together with the unusual situation of having regard to matters after the date of valuation facilitated by the concept of confirming a foresight. However, to effectively rely on the foresight principle, the valuer should clearly state that which is being foreseen and relied upon as at the date of valuation in their valuation report such that it can be clearly identified for confirmation by later events.

The temporal difference between the concepts of highest and best use and of potentiality was explored, with highest and best use having a short-term focus and requiring the constraints of physically possible, legally permissible and financially feasible to be capable of being achieved through the eyes of the hypothetical purchaser in the actual market as at the valuation date. The concept of potentiality has a long-term focus and may be considered, in the context of rezoning, to have four stages, each of which has a differing impact on value. The concepts introduced in this chapter provide the essential foundation principles of statutory valuation that are then applied in the following chapters on statutory valuation.

Chapter 2 considers the practice of statutory valuation including valuation methods for statutory valuation with an emphasis on the comparative method of valuation, together with statutory valuation conventions such as the before and after method and the piecemeal method. This provides an understanding of the valuation methods and conventions that may be preferred for application in the following chapters on statutory valuation.

Statutory valuation for acquisition is considered in Chapters 3, 4, 5 and 6. Chapter 3 introduces the principles of statutory valuation for acquisition, including the statutory framework and process, with a focus on the concept of disregard of the scheme in compensation assessment and the distinction between concepts of value and worth. This provides an insight into concepts of fairness and equity in statutory valuation in the context of the disregard of compulsion.

Chapter 4 explores the market value bases of compensation in acquisition for which statutory valuation may be required, including market value of the interest and the related market value compensation concepts of severance and injurious affection. This provides a broad understanding of the heads of compensation for acquisition on a value basis that may arise in statutory valuation.

Chapter 5 considers non-market value forms of compensation including special value, which is noted to rarely arise, payment for solace arising from loss of home, which is heavily dependent on the exact wording of the facilitating statute and the concept of disturbance for costs and losses suffered as a result of the acquisition. This provides a broad understanding of the heads of compensation for acquisition that are not necessarily on a value basis but that may arise in statutory valuation.

Chapter 6 concludes the chapters considering statutory valuation for acquisition by examining compensation for other interests such as easements and mortgages, mineral rights and profits à prendre, strata title and substrata, businesses and the emerging areas of native title and of indigenous cultural loss. This provides an appreciation of the diversity of interests that may be subject to acquisition and so may require valuation for the assessment of compensation.

Statutory valuation for rating and taxing is considered in Chapters 7 and 8. Chapter 7 introduces the principles of land as the basis for taxation through land tax, stamp duty and rates, considering the general principles of the valuation of land for taxation using the New South Wales statutory framework as an example. This provides a general understanding of the often arcane and artificial basis of statutory valuation upon which Government collects very significant amounts of taxation revenue.

Chapter 8 explores the practice of statutory valuation for rating and taxing through exemptions, concessions and allowances that have evolved over the last century, together with contemporary issues such as mass appraisal, automated valuation models and artificial intelligence, the impact of major disasters on valuation for rating and taxing and the litigation process including the valuer as an expert witness. This provides an insight into the continued evolution of statutory valuation to address issues as they arise over time, highlighting the importance of a principles-based approach to statutory valuation.

Regulatory valuation, considered in the context of the preparation of valuations for inclusion in financial reporting, is considered in Chapters 9 and 10. Chapter 9 explores the principles of regulatory valuation with regard to the inter-related roles of international accounting standards, international financial reporting standards and international valuation standards, focusing on the primacy of fair value in international financial reporting standards. This provides an understanding of the fundamental conceptual differences and bases between statutory valuation and regulatory valuation.

Finally, Chapter 10 concludes regulatory valuation by exploring the practice of valuation approaches and methods relevant for regulatory valuation such as the income approach and the cost approach and the various valuation methods there within. This provides an appreciation of the wider range of valuation approaches and methods available to the valuer for regulatory valuation, reflecting that, unlike statutory valuation, there is no requirement for defence in Court (negligence claims excepted).

References

Baum A, Sams G, Ellis J, Hampson C and Stevens D (2014) *Statutory valuations*, Routledge, Abingdon
Brown, D (2009) *Land acquisition*, Lexis Nexis Butterworths, Chatswood
Graaskamp, JA (1977) *The appraisal of 25 North Pinckney*, Landmark Research, Madison
IVSC (2024) *International valuation standards*, IVSC, London
Jacobs, M (2015) *Law of compulsory land acquisition*, Thomson Reuters, Pyrmont
Parker, D (Ed) (2022) *Principles and practice of property valuation in Australia*, Routledge, Abingdon
Parker, D (2024) *Introduction to property valuation in Australia*, Routledge, Abingdon
RICS (2024) *RICS valuation – global standards (Red Book)*, RICS, London
Rost, RO and Collins, HG (1990) *Land valuation and compensation in Australia*, 3rd Edition, Australian Institute of Valuers and Land Administrators, Deakin
Valuer General (2021) *Review of the impact of rezoning potentiality on land values*, Valuer General, Sydney
Whipple, RTM (2006) *Property valuation and analysis*, Lawbook Co, Sydney

2 Statutory valuation

Practice

2.1 Introduction

This book is an introduction to two very different valuation worlds in Australia – the complex and often arcane national realm of statutory valuation and the dynamic global empire of regulatory valuation. Statutory valuation is different from regulatory valuation, with the former being based on relevant applicable statutes and being capable of challenge on appeal in Court while the latter is based on relevant applicable regulations and incapable of challenge in Court (negligence claims excepted).

As an introductory textbook on statutory and regulatory valuation, this book is a companion to the Australian valuation textbooks, *Introduction to Property Valuation in Australia* (Parker, 2024) and *Principles and Practice of Property Valuation in Australia* (Parker, 2022), which provide a wider analysis of key principles underlying property valuation and current techniques and issues in the practice of property valuation for major sectors of the Australian property market.

Statutory valuation in Australia is most commonly encountered in valuations for rating, taxing and acquisition but may arise wherever a statute prescribes a basis of or approach to valuation. Regulatory valuation is most commonly encountered in valuations for financial statements but may arise wherever a regulation prescribes a basis of or approach to valuation.

Rating and taxing of land raise significant amounts of revenue for State Governments and Councils in Australia, based on statutory valuation. For example, the NSW 2025–26 Budget Statement estimates land tax revenue for 2024–2025 to be $8.2 billion with the City of Brisbane Annual Plan and Budget 2024–2025 estimating proposed rates and utility charges to be $1.51 billion.

Acquisition of land, property and businesses commonly arises for major infrastructure projects and may be by agreement or by compulsion, based on statutory valuation. For example, such major infrastructure projects in Australia currently include the multi-billion-dollar Sydney Metro, Melbourne Suburban Rail Loop and Melbourne to Brisbane Inland Rail projects.

Valuation of land and property holdings based on regulatory valuation arises for all reporting entities in Australia that publish financial statements upon which it is reasonable to expect that users will rely for decision making. For example, at June 2025 there are over 2,000 reporting entities listed on the Australian Stock Exchange which each publish financial statements, with the financial statements for Goodman Group, as at 30 June 2024, showing the fair value of total property assets to be $20.2 billion, based on regulatory valuation.

Focusing on the key issues from a valuer's perspective, the concepts and principles of statutory valuation are introduced in Chapters 1 and 2, with statutory valuation for acquisition considered in Chapters 3, 4, 5 and 6 and statutory valuation for rating and taxing considered in Chapters 7 and 8. Regulatory valuation is then considered in Chapters 9 and 10.

DOI: 10.1201/9781032616957-2

Chapter 1 introduced the principles of statutory valuation, including the concepts of market value, liberal/conservative estimates of value, highest and best use, potentiality and confirmation of a foresight. This provided the essential foundation principles of statutory valuation that are then applied in the following chapters on statutory valuation.

This chapter considers the practice of statutory valuation including valuation methods for statutory valuation with an emphasis on the comparative method of valuation, together with statutory valuation conventions such as the before and after method and the piecemeal method. This provides an understanding of the valuation methods and conventions that may be preferred in the following chapters on statutory valuation.

Statutory valuation for land acquisition is considered in Chapters 3, 4, 5 and 6, supported by an evolving example in Appendix 1 of each chapter. Chapter 3 introduces the principles of statutory valuation for land acquisition, including the statutory framework and process, with a focus on the concept of disregard of the scheme in compensation assessment and the distinction between concepts of value and worth. This provides an insight into concepts of fairness and equity in statutory valuation in the context of the requirement to disregard the compulsory nature of the acquisition.

Chapter 4 explores the market value bases of compensation in acquisition for which statutory valuation may be required, including market value of the interest and the related market value compensation concepts of severance and injurious affection. This provides a broad understanding of the heads of compensation for acquisition on a value basis that may arise in statutory valuation.

Chapter 5 considers non-market value forms of compensation including special value, which is noted to rarely arise, payment for solace arising from loss of the primary place of residence and the concept of disturbance for costs and losses suffered as a result of the acquisition. This provides a broad understanding of the heads of compensation for acquisition that are not necessarily on a value basis but which may arise in statutory valuation.

Chapter 6 concludes the chapters considering statutory valuation for acquisition by examining compensation for other interests such as easements and mortgages, mineral rights and profits à prendre, strata title and substrata, businesses and the emerging statutory valuation areas of native title and indigenous cultural loss. This provides an appreciation of the diversity of interests that may be subject to acquisition and so may require valuation for the assessment of compensation.

Statutory valuation for rating and taxing is considered in Chapters 7 and 8. Chapter 7 introduces the principles of land as the basis for taxation through land tax, stamp duty and rates, considering the general principles of the valuation of land for taxation using the New South Wales statutory framework as an example. This provides a general understanding of the often arcane and artificial basis of statutory valuation upon which Government collects very significant amounts of taxation revenue.

Chapter 8 explores the practice of statutory valuation for rating and taxing through exemptions, concessions and allowances that have evolved over the last century, together with contemporary issues such as mass appraisal, automated valuation models and artificial intelligence, the impact of major disasters on valuation for rating and taxing and the litigation process including the valuer as an expert witness. This provides an insight into the continued evolution of statutory valuation to address issues as they arise over time, highlighting the importance of a principles-based approach to statutory valuation.

Regulatory valuation, considered in the context of the preparation of valuations for inclusion in financial reporting, is considered in Chapters 9 and 10. Chapter 9 explores the principles of regulatory valuation with regard to the inter-related roles of international accounting standards, international financial reporting standards and international valuation standards, focusing on the primacy of fair value in international financial reporting standards. This provides an understanding of the fundamental conceptual differences and bases between statutory valuation and regulatory valuation.

Finally, Chapter 10 concludes regulatory valuation by exploring the practice of valuation approaches and methods relevant for regulatory valuation such as the income approach and the cost approach and the various valuation methods there within. This provides an appreciation of the wider range of valuation approaches and methods available to the valuer for regulatory valuation reflecting that, unlike statutory valuation, there is no requirement for defence in Court (negligence claims excepted).

This book is based on statute, case law and regulations as at March 2025 and those standards and guidance documents published in IVSC (2024) and RICS (2024) and on the API website (accessed January to March 2025). Given their nature, statutes, case law, regulations, standards and guidance documents are dynamic and change continuously. Accordingly, readers should not rely on this book as a current statement of a statute, case law position, regulation, standard or guidance document and should undertake their own enquiries to find the most recent version.

Accordingly, this chapter will now consider:

- the comparative method of valuation;
- alternative acceptable methods of valuation – income approach;
- alternative acceptable methods of valuation – cost approach and
- the before and after method and the piecemeal method of valuation.

2.2 Comparative method of valuation

For statutory valuation purposes, the method of valuation to be adopted is not usually specified, with the choice of method left to the valuer to determine based on suitability. However, while different case law in different jurisdictions has frowned upon certain methods of valuation, all jurisdictions accept the comparative method of valuation:

> It is well established that, if comparable sales are available, the direct comparison of sales evidence approach is the conventional method of valuation (Graham Trilby Pty Ltd v Valuer General [2009] NSW LEC 1087 at 41(Trilby); Redeam Pty Ltd v South Australian Land Commission (1977) 40 LGRA 151 at 156; Riverbank Pty Ltd v Commonwealth (1974) 48 AJLR 483 at 484; Marroun v Roads and Maritime Services [2012] NSWLEC 199 at 196 (Marroun); El Boustani v Minister Administering the Environmental Planning and Assessment Act 1979 [2012] NSWLEC 266 at 23 (El Boustani); Tenstat Pty Limited v Valuer General/Woolworths Limited v Valuer General [2012] NSWLEC 1361 at 35 (Tenstat)).
>
> (*Adams v Valuer General* [2014] NSWLEC 1005 at 28)

as:

> This is because if comparable sales are available they represent direct evidence of the market's valuation of all of the variables that must otherwise be taken into account . . .
>
> (*Jessica Investments Pty Ltd v Valuer General* [2008] NSWLEC1375, 7)

Therefore, for statutory valuation, using the comparative method of valuation is a safe place for a valuer to start. The comparative method is based on the economic principle of price equilibrium, whereby the value of a property interest is indicated by comparison to identical or similar property interests for which price information is available. Effectively, the comparative method seeks to assess the value of a property interest by comparing the subject property interest to recent transactions of similar property interests. While the concept is simple, the application is often challenging.

At the heart of the comparative method of valuation is comparison, such that the level of similarity between the comparable sales and the subject property is fundamental. If the level of similarity is high, the comparative method may be explicit, transparent and reliable. If the level of similarity is low, the level of judgement required in accumulation, analysis, adjustment and application is higher so that the method may be less explicit, less transparent and less reliable.

Caution is required where comparable sales arise in special circumstances, including:

- forced sales, such as by mortgagees or receivers;
- sales due to ill health, financial difficulties, family disputes or a need for a quick sale or
- sales by Councils for the recovery of outstanding rates.

as these circumstances may depress sale prices, or the following special circumstances that may enhance sale prices:

- purchase by an adjoining owner;
- where the property is not on the market and the seller is persuaded by the buyer to sell;
- where the buyer is influenced by family or sentimental reasons;
- acquisition by statutory authorities which may include compensation for other than market value;
- where a tenant is compelled to purchase occupied premises in order to retain business goodwill;
- imprudent or ill-informed parties to a sale or
- sales on unduly liberal terms and conditions.

or sales between closely related parties (such as family members or inter-company sales, being not 'arms-length') or sales with unusual terms and conditions, with care required in determining the weight to be attached to such comparable sales evidence (Rost and Collins, 1990).

However, it may often be the case that, rather than being considered as comparable or not comparable, sales may have a level of comparability or weighting that should be taken into account:

> It is sometimes thought that sales are comparable or not comparable: that is, a binary paradigm should be used to classify sales. In my opinion, this is a flawed approach. Rather there will be gradations of comparability: from identical to irrelevant. As this scale of comparability approaches the irrelevant end, there will be many sales that offer so little assistance that they ought be disregarded. Further, there will be circumstances where there is a sale or sales that are strongly comparable, in which case there will be no need to closely analyse other sales, even though these may be comparable in some way.
>
> (*ISPT Pty Ltd v City of Melbourne* [2007] VCAT 652 at 50–54)

The properties chosen for comparison should be similar to the subject property in terms of location, size, construction type, planning, tenure and so forth, with the comparable transactions being within a period of (ideally) 12 months prior to the date of valuation.

For example, if the subject property is a two-bedroom brick house, sales within the last 12 months of other two-bedroom brick houses in close proximity to the subject property will be most helpful. A sale five years ago of a four-bedroom timber house located three kilometres from the subject property is less likely to be helpful.

In periods of market volatility, the period that is considered recent may be less than 12 months. In rapidly rising or falling markets, values may change significantly in three or six months such that sales from 12 months ago may be of limited assistance.

In the event of the absence of comparable sales data, a Court may accept other valuation methods, as will be considered in Sections 2.3 and 2.4.

2.2.1 Approach to the comparative method of valuation following **Adams v Valuer General** *[2014] NSWLEC 1005*

In *Adams v Valuer General* [2014] NSWLEC 1005, the Court provided guidance on how to undertake a valuation using the comparative method of valuation, such that the valuation may be logically supported by comparable sales evidence and not simply an assertion based on judgement and implicit assumptions. The Court viewed the comparative method of valuation as comprising four sequential steps, being:

- the accumulation step;
- the analysis step;
- the adjustment step and
- the application step.

2.2.1.1 Adams *approach – accumulation step*

The accumulation step is described in *Adams* as follows:

> The accumulation of potentially genuinely comparable sales seeks to identify and establish a pool of relevant comparable sales from which information may be deduced concerning the value of the subject property.
>
> (para 32)

Reflecting the heterogeneity of property, it is preferable at this stage in the process to include as many sales as possible in the sample for consideration, provided each shares similar characteristics to the subject property and is therefore a relevant comparable sale, as the sample size is likely to be diminished during the course of undertaking the following steps.

For the effective application of the comparative method of valuation, extensive, detailed and accurate information about comparable sales is essential. In the past, valuation firms maintained meticulous paper-based file records of property sales, which provided a significant source of competitive advantage. Today, it is common to rely on online data sources, with, in an Australian context, Domain, RP Data and State Valuer General websites being popular.

While there is no maximum or minimum ideal number of comparable sales, several are preferable. Unless directly comparable, one or two may be too few to form a view of the market, whereas more than six may be challenging to appropriately comprehend, as generally, up to seven separate entities can be considered by the human brain simultaneously. However, a paired comparison of two sales may be helpful to identify the effect of a particular characteristic or the impact of time and market changes.

To fully understand the context of comparable sales, it may be necessary to contact the vendor's agent or other valuers, reinforcing the importance of networks and personal contacts for the valuer to verify sales and gather more information.

Comparable sales used for statutory valuation purposes should be based on binding contracts of sale between arms-length parties following voluntary negotiation of price and after a reasonable period of marketing. There is extensive case law concerning the acceptability as comparable of other forms of price evidence such as asking prices, potential sales, conditional sales, offers to sell/purchase, options and so forth in different circumstances, with readers are referred to Hyam (2014), Brown (2009) and Jacobs (2015) for greater detail.

The relevant date for a comparable sale is the date of contract, not the date of settlement, as the date of contract is evidence of market conditions at that date, while market conditions may have changed by the date of settlement.

Brown (2009, page 232) considers, at length, whether sales of land to a resuming authority may be considered as comparable sales evidence. The author notes, with supporting case law, that historic cases considered such sales to be unreliable but that most resumption statutes today include the concept of market value for the land acquired, such that the question now is not one of admissibility but one of the weight to be given to such transactions. Consideration of such sales may arise in two situations: where there are no other sales (such that the sales to the resuming authority fill a void that would otherwise exist) or where they are amongst a number of sales (and so may be confirmatory, with the issue then going to the weight to be attached to the sales).

2.2.1.2 Adams *approach – analysis step*

The analysis step is described in *Adams* as follows:

> The analysis of potentially genuinely comparable sales provides a common basis of measurement by seeking to convert all potentially comparable sales to a common basis of expression such as a unitary rate (rate per square metre, rate per hectare, etc), improved or unimproved (through allowance for the absence or existence of improvements, etc) and so forth.
>
> (38)

The unitary rate should be that commonly used in the market for that type of property (such as a rate psm, rate per hectare, rate per lot and so forth) and it is important that the same unitary rate is adopted for all comparable sales in the sample to permit comparison on a common basis.

While land sales may be analysed to a common basis of rate per square metre or rate per hectare, residential property sales may not require analysis except where they lack a common basis, such as being partially improved, improved with a dilapidated or derelict dwelling or improved differently (such as with or without a garage, car port, swimming pool and so forth).

2.2.1.3 Adams *approach – adjustment step*

The adjustment step is described in *Adams* as follows:

> 40. The adjustment of potentially genuinely comparable sales acknowledges the fact that no two properties are ever identical and seeks to convert those potentially comparable

sales to a hypothetical expression of value as a unitary rate in the context of the subject property through the reflection of differences (such as size, location, use, date, etc) between the respective potentially comparable sales and the subject property.

41. Because properties are never identical, explicit and/or implicit adjustment for differences is obviously necessary but caution is required through making as few adjustments as possible, in a consistent manner, to ensure the reliability of the comparable sale when related to the subject property, with too much adjustment potentially rendering the comparable sale unsafe to use.

42. Therefore, as a matter of general valuation principle, fewer adjustments may be preferred to more adjustments and smaller adjustments may be preferred to larger adjustments in rendering comparable sales safe for use.

Very significantly, the Court noted that *Caution in adjustment is required, as too much adjustment renders the use of comparables unsafe* (para 44), with the notion of safety indicating that no comparable transaction is necessarily irrelevant, though care is required with the level of adjustment, as too much adjustment may render the comparable transaction of limited relevance (considered further at para 57).

Further, smaller adjustments are preferred to larger adjustments and explicit adjustments are preferred to implicit adjustments, such that particular caution is required with large implicit adjustments (para 46).

As Croft J stated in *Challenger Property Asset Management Pty Ltd v Stonnington City Council* [2011] VSC 184:

> Care, however, needs to be exercised with respect to the process of adjusting for comparability, for two reasons. First, the process of adjustment has an illusion of precision which tends to obscure the reality that the process is a matter of expert judgment; an art and not a science. Secondly, it is clear the question whether a sale can be relied upon as comparable is a question of fact and degree and, generally speaking, if the adjustments required to achieve comparability are very extensive, one might be more inclined to the view that the extent of adjustments required indicates that the sales are not, in reality, comparable.

Having accumulated a pool of potentially genuine comparable sales and analysed each to a common unitary rate basis, the comparable sales require adjustment for differences in those aspects that are of relevance to the property type.

Importantly, the adjustment process is prospective to derive an assessment of value rather than retrospective to support an assessment of value previously formed (*Adams v Valuer General* [2014] NSWLEC 1005 at 55). For most properties, adjustment for differences may include land area, planning (zoning, development constraints, etc.) and tenure, with residential property transactions potentially also requiring adjustment for location and views, and office letting transactions potentially also requiring adjustment for floor level and aspect.

Rost and Collins (1990, page 89) suggest adjustment for the following:

- location – with greater reliance placed on comparable sales closest to the subject property than on sales more distant;
- dates of sale and date of valuation – with greater reliance placed on comparable sales at or immediately prior to the date of valuation;

- physical characteristics – with greater reliance based on similarities in area, dimensions, shape, aspect and elevation, surface contours and topography and nature of soil, rainfall and climatic features for rural lands;
- amenities and services – with greater reliance placed on proximity, relative to the subject, to facilities for education, worship, shopping and social life and similarities in services such as roads, public transport, water, sewerage, electricity, gas and telephone;
- title and tenure – which should preferably be the same as the subject property and
- land use and potential utility – being the highest and best use or potentiality.

Generally, up to seven separate entities can be considered by the human brain simultaneously. Therefore, a matrix plotting each comparable sale as columns and around seven entities for adjustment as rows may be capable of comprehension.

Each cell in such a matrix will contain an adjustment that reflects a judgement of relativity by the valuer. While there is no prescribed basis for adjustment, with different valuers preferring different bases, commonly adopted approaches include:

- use of a percentage;
- use of a descriptor such as superior, comparable or inferior;
- use of upward and downwards arrows;
- use of ticks and crosses and
- use of plus/minus signs.

These approaches must be consistently applied in an explicit, transparent and logical manner.

If adjustment by percentages is adopted, the total adjustment should be no more than +/-30% for the comparable sale to be safe for use, and adjustments should follow a common pattern (for example, 2.5%, 5%, 7.5%, 10% or 2%, 4%, 6%, 8%) when judgementally based, with adjustments such as +1.5% or −6.33% requiring clear supporting evidence. An exception to such rounding is for percentage adjustments to reflect market movement over time, which may be derived from a paired sales analysis or a proprietorial database (such as Domain or RP Data) and be rounded to two decimal places.

Importantly, while the matrix approach affords transparency and fosters logical relativity of adjustments, valuer judgement is still required. As Sheahan J stated in *Marroun v Roads and Maritime Services* [2012] NSWLEC 199:

> A transparent process of explicit adjustment, leading to an explicable assessment of value, is to be preferred to an opaque process of implicit adjustment, leading to an assertion of value.
>
> (206)

2.2.1.4 Adams *approach – application step*

The application step is described in *Adams* as follows:

> The application of those potentially genuinely comparable sales to the subject property seeks to determine the value of the subject property through a consideration of the relevance (such as being limited, indirect or direct) of the unitary rate derived from those adjusted comparable sales relative to the subject.
>
> (para 56)

and:

> While all comparable sales evidence may be considered relevant and so cannot be disregarded, the level of relevance of different comparable sales to the subject property may vary leading to the valuer attributing differing weight to different comparable sales.
>
> (para 57)

Having accumulated a pool of potentially genuine comparable sales, analysed each to a common unitary rate basis, and then adjusted each to a hypothetical expression of value as a unitary rate in the context of the subject property through the reflection of differences, the adjusted comparable sales require application to the subject property through a consideration of their individual relevance and the resulting weight to be given to each.

As Sugerman J stated in *Best v Housing Commission of NSW* (1949) 17 LGR (NSW) 129 at 130:

> The correct approach is to assign the subject land by comparison to its proper place in the scale of values disclosed by the sales proved.

While no sale is irrelevant, those with the greatest similarity to the subject may be rated as of direct relevance, those with lower similarity may be rated as of indirect relevance, and those of limited comparability may be rated as of limited relevance. The application step may give greatest weight to those comparable sales of direct relevance in determining the range of value for the subject property, which should then broadly accord with those comparable sales of indirect or limited relevance to which less weight is attributed.

Significantly, the adjusted comparable sales should not be averaged (*Daandine Pastoral Co Pty Ltd v Commissioner of Land Tax* (1943) 7 *The Valuer* 299) (Rost and Collins, 1990) but should be individually applied by the valuer to the subject property using judgement.

An example of the implementation of the four steps identified in *Adams* to the valuation of a residential property is provided in the following section.

2.2.2 *Example of the comparative method of valuation following* **Adams v Valuer General** *[2014] NSWLEC 1005*

You are instructed to assess the market value of a two-bedroom house with a double garage on a 500 sqm block of land at a valuation date of 30 June 2025. The exterior of the house is of timber construction and the interior is in good condition. The house is in an outer suburb, being located close to schools, train station and shops.

2.2.2.1 *Example – accumulation step*

Your search of internal records, relevant databases and websites identifies six potential genuinely comparable sales:

Sale 1

The sale of a two-bedroom house with a single garage on a 500 sqm block of land, three months before the valuation date, for $950,000. The exterior of the house is of brick construction and the interior is in fair condition. The house is in the same outer suburb, being within 200 m of the subject property, close to schools, train station and shops.

Sale 2

The sale of a two-bedroom house with no garage on a 400 sqm block of land, four months before the valuation date, for $850,000. The exterior of the house is of timber construction and the interior is in good condition. The house is in the same outer suburb, being located around 1 km from schools, train station and shops.

Sale 3

The sale of a three-bedroom house with a double garage on an 800 sqm block of land, two months before the valuation date, for $1,150,000. The exterior of the house is of brick construction and the interior is in very good condition. The house is in the same outer suburb, being located around 500 m from schools, train station and shops.

Sale 4

The sale of a two-bedroom house with a double garage on a 600 sqm block of land, six months before the valuation date, for $990,000. The exterior of the house is of timber construction and the interior is in good condition. The house is in the same outer suburb, being located close to schools, train station and shops.

Sale 5

The sale of a four-bedroom house with no garage on a 750 sqm block of land, one month before the valuation date, for $1,100,000. The exterior of the house is of brick construction and the interior is in very good condition. The house is in the same outer suburb, being located about 750 m from schools, train station and shops.

Sale 6

The sale of a five-bedroom house with a double garage on a 1000 sqm block of land, six months before the valuation date, for $1,225,000. The exterior of the house is of brick construction and the interior is in very good condition. The house is in an adjacent outer suburb, being located about 2 km from schools, train station and shops.

2.2.2.2 Example – analysis step

The analysis step seeks to identify a unitary rate that is commonly used in the market for that type of property. It is important that the same unitary rate is adopted for all comparable sales in the sample to permit comparison on a common basis.

In this example, the accumulated comparable sales are all improved residential properties, which provides a common basis of measurement. However, unlike the subject property and other comparable sales, Sale 2 and Sale 5 do not have a garage. Further, the comparable sales took place between one and six months before the valuation date.

Issues such as the absence of a garage and the impact of market changes over time may be addressed either in the analysis step or in the adjustment step. For the purposes of illustration in this example, the absence of a garage in Sale 2 and Sale 5 will be addressed in the analysis step by the addition of $50,000 to the sale price of each, representing the valuer's understanding of the cost of building a double garage in the relevant suburb:

Sale 2	Sale price	$850,000
	+ garage	$50,000
	Analysed sale price	$900,000
Sale 5	Sale price	$1,100,000
	+ garage	$50,000
	Analysed sale price	$1,150,000

Adopting an analysed sale price of $900,000 for Sale 2 and $1,150,000 for Sale 5 provides a common basis of expression for all the comparable sales for use in the adjustment and application steps.

The impact of market changes over time may be addressed by increasing/decreasing the comparable sale price in the analysis step or in the adjustment step. Such an increase/decrease should be based on market evidence, such as from the analysis of paired sales or from data published by a reputable agency, research group or proprietorial database (such as Domain or RP Data).

For the purposes of this example, it is assumed that there has been no market movement over the six-month period leading to the valuation date, so no amendment to the sale prices of the comparable properties is required.

2.2.2.3 Example – adjustment step

The adjustment step is described in *Adams* as being the reflection of differences (such as size, location, use, date and so forth) between the respective potentially comparable sales and the subject property, within the following guidelines:

- fewer adjustments may be preferred to more adjustments;
- smaller adjustments may be preferred to larger adjustments and
- explicit adjustments may be preferred to implicit adjustments.

Adjustments may be a percentage, a description (such as superior, comparable, inferior), arrows, ticks/crosses, plus/minus or other form of comparison but must be consistently adopted by the valuer and must be transparent and explicit.

For the purposes of this example, percentage adjustments will be adopted using a matrix plotting each comparable sale in columns and around six entities for adjustment in rows for ease of comprehension. Care is required to ensure that the percentage adjustments are consistently applied, following a common pattern (for example, 2.5%, 5%, 7.5%, 10% or 2%, 4%, 6%, 8%) and summing to not more than +/–30% if the comparable sale is to be safe for use.

2.2.2.3.1 EXAMPLE – ADJUSTMENT STEP OVERVIEW

Before applying the adjustment factor, it is helpful to overview the similarities and differences between the subject property and the comparable sales:

Sale 1

While otherwise comparable, the brick construction is superior, with the location and fair internal condition being inferior to the subject property. The relativity of the differences will determine if the subject property would be expected to sell for a lower price or higher price than Sale 1.

Sale 2

While otherwise comparable, the smaller block size and location are inferior to the subject property. Therefore, the subject property would be expected to sell for a higher price than Sale 2.

Sale 3

While otherwise comparable, the location is inferior to the subject property, but the number of bedrooms, block size, brick construction and very good internal condition are superior to the subject property. Therefore, the subject property would be expected to sell for a lower price than Sale 3.

Sale 4

While otherwise comparable, the larger block size is superior to the subject property. Therefore, the subject property would be expected to sell for a lower price than Sale 4.

Sale 5

While otherwise comparable, the location is inferior, but the number of bedrooms, block size, brick construction and very good internal condition are superior to the subject property. Therefore, the subject property would be expected to sell for a lower price than Sale 5.

Sale 6

While otherwise comparable, the location is significantly inferior, but the larger number of bedrooms, block size, brick construction and very good internal condition are superior to the subject property. The relativity of the differences will determine if the subject property would be expected to sell for a lower price or higher price than Sale 6.

2.2.2.3.2 EXAMPLE – ADJUSTMENT STEP MATRIX

Adopting an adjustment basis of 2.5% increments (not because accuracy to 0.5% can be assured but reflecting a common pattern of incremental increases), the above inferior and superior aspects of the comparable sales to the subject property may be quantified as shown in Table 2.1.

As the adjustment matrix shows, the greater the difference between the comparable sale and the subject property, the greater the level of adjustment, with adjustments being applied consistently across each of the comparable sales. Further, none of the adjustments sum to more than 30% for a given comparable sale, indicating that the accumulated comparable sales are safe for use.

The issues of inferiority and superiority should be viewed through the lens of making the comparable sale less or more valuable than the subject property. For example, a larger block

Table 2.1 Adjustment matrix

	Sale 1	*Sale 2*	*Sale 3*	*Sale 4*	*Sale 5*	*Sale 6*
Location	+2.5%	+7.5%	+5%	0%	+7.5%	+10%
Land size	0%	+2.5%	–5%	−2.5%	–5%	–10%
Bedrooms	0%	0%	–5%	0%	−7.5%	–10%
Internal condition	+2.5%	0%	–5%	0%	–5%	–5%
External construction	–5%	0%	–5%	0%	–5%	–5%
Total	0%	+10%	–15%	−2.5%	–15%	–20%

Source: Author

Table 2.2 Adjusted sale price matrix

	Sale 1	*Sale 2*	*Sale 3*	*Sale 4*	*Sale 5*	*Sale 6*
Analysed sale price	$950,000	$900,000	$1,150,000	$990,000	$1,150,000	$1,225,000
% Adjustment	0%	+10%	–15%	−2.5%	–15%	–20%
Adjusted sale price	$950,000	$990,000	$977,500	$965,250	$977,500	$980,000

Source: Author

size will make Sale 6 more valuable than the subject property (adjustment to Sale 6 price −10%) whereas fair internal condition will make Sale 1 less valuable than the subject property (adjustment to Sale 1 price +2.5%).

The adjustments from the adjustment matrix may be applied to the analysed sale prices as shown in Table 2.2.

Therefore, the adjusted comparable sale prices for use in the application step range from $950,000 to $990,000.

2.2.2.4 Example – application step

The application step:

> . . . seeks to determine the value of the subject property through a consideration of the relevance (such as being limited, indirect or direct) of the unitary rate derived from those adjusted comparable sales relative to the subject.
>
> (*Adams*, para 56)

Further:

> While all comparable sales evidence may be considered relevant and so cannot be disregarded, the level of relevance of different comparable sales to the subject property may vary leading to the valuer attributing differing weight to different comparable sales.
>
> (*Adams*, para 57)

Therefore, the adjusted comparable sales require application to the subject property through a consideration of their individual relevance and the resulting weight to be given to each. While no sale is irrelevant, those with the greatest similarity to the subject property may be rated as of direct relevance, those with lower similarity may be rated as of indirect relevance, and those of limited comparability may be rated as of limited relevance.

The application step may give greatest weight to those comparable sales of direct relevance in determining the range of value for the subject property, which should then be broadly in alignment with those comparable sales of indirect or limited relevance to which less weight is attributed.

Having regard to their similarity to the subject property, the level of percentage adjustment and the resulting adjusted comparable sale prices, the following levels of relevance may be proposed:

Direct relevance	Sale 1, Sale 4
Indirect relevance	Sale 2, Sale 3, Sale 5
Limited relevance	Sale 6

Giving the greatest weight to those comparable sales of direct relevance, the range of value for the subject property is $950,000 to $965,250, from which the market value of the subject property may be assessed by valuer judgement (not averaging) at $960,000.

As a final quality control check, the valuer should step back and consider the assessment of $960,000 to ensure that it is in alignment with the comparable sales evidence – effectively asking the fundamental question 'does this valuation make sense?'.

2.2.3 Summary – the comparative method of valuation following **Adams v Valuer General** *[2014] NSWLEC 1005*

While, for statutory valuation purposes, the method of valuation to be adopted is not usually specified, with the choice of method left to the valuer to determine based on suitability, adoption of the comparative method of valuation is accepted as the preferred approach in most jurisdictions.

However, adoption of the comparative method of valuation assumes the existence of comparable sales for accumulation, analysis, adjustment and application following the guidance of the NSW Land and Environment Court in *Adams*. While the comparative method of valuation may be most helpful for residential property, small commercial property, rural property and land, other types of property may lack comparable sales and so require an alternative method of valuation.

2.3 Alternative acceptable methods of valuation – income approach

For statutory valuation purposes, the method of valuation to be adopted is not usually specified, with the choice of method left to the valuer to determine based on suitability. While different case law in different jurisdictions has frowned upon certain methods of valuation, all jurisdictions accept the comparative method of valuation. However, in the absence of comparable sales or where a check method of valuation may be required, the Courts have considered a range of alternative acceptable methods of valuation, including methods within the income approach to valuation, considered below, and methods within the cost approach to valuation, considered in Section 2.4.

Brown (2009) notes that, in *Bronzel v State Planning Authority* (1979) 44 LGRA 34 at 38, it was held that the Court was not disposed to reject any method of valuation adopted by a valuer:

> . . . this court should be slow to reject any method that, in expert hands, is capable of yielding a result within bounds that are not unreasonable.
>
> (38)

Further, the usual basis for seeking to adopt alternative valuation methods is the statement by Callinan J in the High Court in *Boland v Yates Property Corporation Pty Ltd* (1999) 167 ALR 575 that:

> There is no legal principle that purports to, or could, close for all times the categories of methods of valuation which might be acceptable in a particular case . . . Valuation practice is, however, like legal practice an evolving discipline.
>
> (280)

Parker (2024) notes that contemporary valuation theory and practice recognise three valuation approaches, within each of which are a range of valuation methods:

- market approach – including the:
  - comparative method of valuation considered in section 2.2;
- income approach – including the:
  - capitalisation of income method;
  - the profits method and
  - the discounted cash flow method and
- cost approach – including the:
  - replacement cost method – including the:
    - depreciated replacement cost method/contractor's method;
    - Australian summation method and
    - reinstatement cost method;
  - reproduction cost method;
  - international summation method and
  - residual or hypothetical development method.

Brown (2009) notes the following four alternative valuation methods have been accepted by the Courts for statutory valuation purposes:

- capitalisation of income method;
- discounted cash flow method;
- summation method and
- development possibilities (hypothetical development method or residual method).

These methods will be considered below through the income approach and the cost approach, consistent with contemporary valuation theory and practice.

2.3.1 *Valuation methods within the income approach*

The income approach to valuation is grounded in economic theory, being based on the economic principle of anticipation of benefits. Within the income approach, commonly adopted valuation methods include the capitalisation of income method, the profits method and the discounted cash flow method, which are briefly considered, respectively, in the following sections, with further detail and examples provided in Chapter 10.

2.3.2 Capitalisation of income method

Commonly used in non-statutory valuation for the assessment of the value of income-producing property, the capitalisation of income method involves the assessment of the current market rental value net of outgoings of the property, which is then multiplied by the relevant capitalisation rate, both being sourced from comparable rental agreements and sale transactions.

Brown (2009) notes that the capitalisation of income method has been accepted by the Courts as appropriate where the nature of the land is such that a hypothetical, prudent purchaser would wish to purchase it for the purpose of investment or income, such as an office block (*Abbey Orchard Property Investments Pty Ltd v Sydney City Council* (1978) 37 LGRA 230), a butcher's shop (*Doherty v Commissioner of Highways No 2* (1974) 7 SASR 57) or a petrol service station (*Hill v Commissioner of Highways* [1966] SASR 316).

The capitalisation of income method may also be used for the valuation of leasehold interests where the passing rent is below the market rent, creating a profit rent. In *Riverstone Meat Co Ltd v Sydney City Council* (1956) 1 LGRA 216 at 219, the Court accepted that the usual method of valuing a leasehold interest is by the capitalisation of the profit rent for the unexpired portion of the term.

Historic cases, from periods prior to the widespread use of the capitalisation of income method, consider the role of depreciation, future vacancies and gilt-edged securities. However, as the capitalisation of income method is now in common use in contemporary valuation practice, the basic format and components are generally agreed within the valuation profession, as detailed in chapter 7 of Parker (2024).

Usually, the principal variables for dispute in the application of the capitalisation of income method are the current market rental value net of outgoings and the capitalisation rate, which are considered further in the example within Chapter 10.

2.3.2.1 Profits method

Commonly used in non-statutory valuation for the assessment of the value of either properties which are inextricably linked to the business operating therein or of a business as a going concern, the profits method is similar to the capitalisation of income method in that it involves the assessment of a form of underlying profit, which is then multiplied by the relevant capitalisation rate, both being sourced from transactions of comparable business properties or businesses.

Types of property which are inextricably linked to the business operating therein are usually specifically adapted for the use of the business and unlikely to be capable of use for other purposes, such as hotels, pubs, bars, restaurants, nightclubs, petrol filling stations, casinos, cinemas, theatres and various other forms of leisure property. Other types of property, such as car parks, garden centres, caravan parks, self-storage and purpose-built student housing, may exhibit less specific use and adaptation constraints and may suit valuation by other methods.

The valuation of a property specifically adapted for the use of a business should be distinguished from the valuation of a business as a going concern within a generic purpose property such as a shop. For a property valuer, the valuation of a business as a going concern may be likely to arise in the context of a small business holding a leasehold or licence interest in a property (such as a newsagent operating a news agency within a leased shop property), as a large business as a going concern is likely to be valued by a business valuer or an accountant.

The potential for dispute is significant, as the profits method requires the assessment of a form of underlying profit (which is the result of the assessment of a wide range of input variables) and the assessment of the relevant capitalisation rate, each of which is considered further in the example within Chapter 10.

2.3.3 Discounted cash flow method

Commonly used in non-statutory valuation for the assessment of the value of income-producing property, the discounted cash flow method involves the discounting of a future cash flow to a present value amount as at the date of valuation at a relevant discount rate.

Concerning discounted cash flow, in *Albany v Commonwealth* (1976) 60 LGRA 287 at 297, Jacobs J famously stated:

> I should now say that I am not satisfied that this could be an acceptable method of valuation in the present case.

Disregarding the qualifier 'in the present case', the Australian valuation profession relied on *Albany* as precedent to assert that the discounted cash flow method was not a method of valuation for the next two decades. It was not until the use of the discounted cash flow method of valuation by AREITs and institutional investors became widespread in the late 1980s that the Australian valuation profession accepted it as a method of valuation in the early 1990s.

Despite the widespread use of the discounted cash flow method by the property industry and the property profession today, the Courts have only adopted the method with reservations, citing particular concerns about the number of assumptions required (see, e.g. *Phillip Street Properties Pty Ltd v Queensland* (1999) 21 QLCR 24, *Collex Pty Ltd v Roads and Traffic Authority of NSW* [2006] NSWLEC 579, *McDonald's Australia Ltd v Transport Infrastructure Development Corporation* [2006] NSWLEC 796).

While the Courts are willing to accept the discounted cash flow method of valuation in principle, its appropriate use is case specific and heavily dependent on a sufficient evidentiary basis for the assumptions made therein. The discounted cash flow method of valuation is considered in greater detail with an example in Chapter 10.

2.3.4 Alternative acceptable methods of valuation – income approach

For clarity, the comparative method of valuation is the preferred method of valuation for statutory valuation purposes, being acceptable to Courts in all jurisdictions. Accordingly, alternative valuation methods should only be considered in the absence of comparable sales or where a check method of valuation may be required.

Alternative valuation methods may include methods within the income approach to valuation and the cost approach to valuation. However, there are significant gaps between contemporary valuation theory and practice and the valuation methods accepted by the Courts over the last century, such that care is required in the interpretation and application of historic case law.

Within the income approach to valuation, the Courts have considered and, in some cases, accepted the capitalisation of income method (and the related profits method) and the discounted cash flow method, with further details and examples of each being provided in Chapter 10. It should not, however, be automatically assumed that such methods will be accepted by the Courts in every case, as the acceptability of such methods is heavily dependent on the facts of the specific case.

2.4 Alternative acceptable methods of valuation – cost approach

For statutory valuation purposes, the method of valuation to be adopted is not usually specified, with the choice of method left to the valuer to determine based on suitability. While different case law in different jurisdictions has frowned upon certain methods of valuation, all jurisdictions accept the comparative method of valuation. However, in the absence of comparable sales or where a check method of valuation may be required, the Courts have considered a range of alternative acceptable methods of valuation, including methods within the income approach to valuation, considered above, and within the cost approach to valuation, considered below.

The cost approach to valuation is grounded in economic theory, being based on the economic principle that a buyer will pay no more than the cost to obtain an asset of equal utility, whether by purchase or construction.

Parker (2024) notes that contemporary valuation theory and practice recognise the following valuation methods within the cost approach:

- replacement cost method – including the:
  - depreciated replacement cost method/contractor's method;
  - Australian summation method and
  - reinstatement cost method;
- reproduction cost method;
- international summation method and
- residual or hypothetical development method.

A review of Jacobs (2015) and Brown (2009) suggests that the Courts have, historically, focused on the cost approach through notions of replacement cost, variously expressed as the replacement cost method or the (Australian) summation method (which, it should be noted, differs from the international version of the summation method) and the hypothetical development method or residual method. Hence, the varying use of terminology by the various Courts makes it challenging to apply some historic case decisions on the cost approach in a contemporary context.

As the Courts do not appear to have considered the cost approach in the context of the reinstatement cost method, the reproduction cost method or the international summation method, the following will focus on the replacement cost method, the Australian summation method and the hypothetical development method or residual method.

It should be noted, however, that the cost approach is fundamentally flawed by its combination of notions of cost and notions of value. Wyatt (2023) validly cautions against the careless intermingling of concepts of cost and concepts of value, with cost being a production-related concept and value being an exchange- or use-related concept. Parker (2024) observes that mixing notions of value and notions of cost is akin to trying to mix oil and water. Notions of value and notions of cost do not sit well together, and the result is a curious expression that is neither a pure expression of value nor a pure expression of cost.

2.4.1 Replacement cost method of valuation

Jacobs (2015) defines the *cost method* as follows:

> The cost method entails the determination of the replacement cost of the buildings on the resumed property, deducting therefrom a fair rate of depreciation and adding back the market value of the land.
>
> (page 473)

Further, Jacobs (2015) notes that replacement cost, being a calculation of the cost of improvements, may involve the use of one or more of three types of costing:

- dollars per square foot (*R v Spencer* [1940] 1 DLR 575);
- unit in place cost (*R v Supertest Petroleum Corp* [1954] 3 DLR 245) and/or
- the cost as determined by a quantity surveyor (*R v Sisters of Charity of Providence* [1952] 3 DLR 358).

Parker (2024) notes that, in Australia today, a rate per square metre for use in the calculation of replacement cost may commonly be sourced from a cost guide such as Rawlinsons Australian Cost Handbook (2024).

Concerning depreciation, Jacobs (2015) identifies two accepted methods of determining depreciation, being:

- the straight line method, with a constant deduction of a certain percentage each year calculated over the estimated lifetime of the improvements or
- the sinking fund method, whereby a fund is created during the lifetime of the improvements in order to enable them to be replaced at the end of their estimated lifetime.

Parker (2024) notes that, in Australia today, depreciation should reflect physical, functional and economic obsolescence, with typical depreciation profiles including:

- straight line, being the same proportion of the original cost deducted for each period of the estimated life of the asset; or
- diminishing value, being the deduction of a constant percentage rate from the cost at the start of the previous period over the estimated life of the asset; or
- S-curve, being the deduction of different percentage rates for each period over the estimated life of the asset, such as higher rates in the initial period reducing over the estimated life of the asset.

Further details and an example of the replacement cost method of valuation are provided in Chapter 10.

2.4.1.1 Australian summation method of valuation

In *Marcus Clark & Co Ltd v Commissioner for Railways* (1949) 29 LVR 98, Sugerman J noted:

> In broad terms, the 'summation method' aims at arriving at the value of premises by adding together the unimproved value of the land and the value of the building.

Brown (2009) notes two approaches that may be contended to be the summation method of valuation in Australia:

- . . . the values of the constituent parts of a property are estimated separately and such values are added together to arrive at the total value of the land (page 250) – citing *Vains v Gosford Shire Council* (1976) 24 *The Valuer* 756 where the value of four land types (creek flats, banks/old creek terraces, lightly timbered hills and heavily timbered hills) was summed (being similar, conceptually, to the international summation method); and

- . . . the addition of the various components of a property, that is, land and buildings, usually on a replacement cost less deprecation basis or direct comparison with the improvement value analysed from sales (page 251) (being the valuation method described as summation by Rost and Collins (1990, page 106)).

Further, Valuer General (2024) provides the following example of the summation method:

> You may, for example, determine the market value of an improved rural property comprising a mix of land classes by separately considering the value of the land and the improvements.
>
> The value of the different classes of land may be determined by direct comparison with sales of other farm land reflecting those different classes.
>
> The farm improvements may be valued preferably by calculating their added value to the land referable to analysed market evidence. In some instances these improvements may be valued by deprecation of the cost new.
>
> Where depreciation is applied to the cost to build the improvements, the rate of depreciation must be rationalised.
>
> Take care when assessing the added value of unique or unusual improvements as they may not represent the highest and best use of the land. Even newly built improvements may not add the same level of value as the cost to build them.
>
> (page 28)

Accordingly, the Australian summation method of valuation may be considered to be the Australian version of the replacement cost method of valuation, for which further details and an example are provided in Chapter 10.

2.4.2 Hypothetical development method or residual method of valuation

The hypothetical development method was summarised by Sugerman J in *AG Robertson Ltd v Valuer General* (1952) 18 LGR (NSW) 261 at 262 as follows:

> (the method) erects a hypothetical building upon the subject land, capitalises the anticipated return therefrom, and subtracts the estimated building cost from the capitalisation, the balance being treated as the (value of the site)

While the hypothetical development method has been accepted by the Courts in a number of matters (see *Gwynvill Properties Pty Ltd v Commissioner for Main Roads* (1983) 50 LGRA 322 and, for a recent example, *AMP Macquarie Pty Ltd v Valuer General* [2022] NSWLEC 114), the Courts have expressed reservations about the number of inter-acting variables and assumptions required, with minor differences in such assumptions and variables resulting in major differences in the assessment of value (see, e.g. *Co-ordinated Resources Pty Ltd v Valuer General* (1983) 27 *The Valuer* 779, *Western Australian Planning Commission v Arcus Shopfitters Pty Ltd* [2003] WASCA 295).

Indeed, Rost and Collins (1990) succinctly note:

> Because of the many estimates and assumptions which must be made, the use of this method can be perilous.
>
> (page 65)

as may be seen from the significant number of valuation negligence cases arising from its use (see, e.g. *Mount Banking Corporation Limited v Brian Cooper & Co* [1992] 35 EG 23 and *Allied Trust Bank Limited v Edward Symmons & Partners* [1993] EGCS 163).

Despite such misgivings, the hypothetical development method has frequently been accepted by the Courts for the valuation of land ripe for subdivision but not yet subdivided (*Closer Settlement Ltd v Minister* (1942) 17 LGR (NSW) 62), vacant land and property for development or redevelopment.

As Parker (2024) notes, the hypothetical development method of valuation may be a static approach, being a single period statement of the present value of each input (often referred to as a residual valuation), or a dynamic approach (being a multi-period cash flow approach) which is usually calculated in a spreadsheet or proprietorial software system. The critical difference between the static and dynamic approaches concerns the cost of finance, with the static approach relying on heuristic assumptions while the dynamic approach sets out the cash flow in periods over the development, allowing a more accurate assessment of the cost of finance and the resulting amount available for land acquisition.

The key inputs for the hypothetical development method of valuation are the value of the completed project (land subdivision, property development), the cost of development (building costs, professional fees, finance) and an allowance for profit and risk. The assessment of the value of the completed project and the cost of development are dependent on a wide range of subjectively determined variables, each of which may be subject to change, impacting the final assessment of value. Concerning the profit and risk allowance, in *Albany v Commonwealth* (1976) 60 LGRA 287 the Court stated that the determination of a:

> [p]rofit and risk factor is calculated as a matter of judgment based upon comparable developments.

Further details and an example of the hypothetical development method and the residual method of valuation are provided in Chapter 10.

2.4.3 Alternative acceptable methods of valuation – cost approach

For clarity, the comparative method of valuation is the preferred method of valuation for statutory valuation purposes, being acceptable to the Courts in all jurisdictions. Accordingly, alternative valuation methods should only be considered in the absence of comparable sales or where a check method of valuation may be required.

Alternative valuation methods may include methods within the income approach to valuation and the cost approach to valuation. However, there are significant gaps between contemporary valuation theory and practice and the valuation methods accepted by the Courts over the last century, such that care is required in the interpretation and application of historic case law.

Within the cost approach to valuation, the Courts have considered and, in some cases, accepted the replacement cost method (and the related Australian summation method) and the hypothetical development method (and the related residual method) with further details and examples of each being provided in Chapter 10. It should not, however, be automatically assumed that such methods will be accepted by the Courts in every case, as the acceptability of such methods is heavily dependent on the facts of the specific case.

2.5 Before and after method and piecemeal method of valuation

The before and after method and the piecemeal method of valuation are only used for valuations for compulsory acquisition purposes within statutory valuation, usually as a result of either injurious affection and/or severance, which will be considered in Chapter 4.

As the names suggest, compensation for injurious affection follows from the suffering of an injury by a landholding (such as the acquisition of land for a sewage facility which detrimentally impacts the landholder's retained land), and compensation for severance follows from the severing of a landholding (such as the acquisition of land for a road through the middle of an agricultural property leaving two separate parcels of land).

In the case of compensation for both injurious affection and severance, only part of a landholding will be compulsorily acquired, with the balance continuing to be held by the landowner.

2.5.1 Before and after method of valuation

The before and after method has been accepted by the Courts in numerous cases (see, for example, *Roads and Traffic Authority of NSW v Damjanovic (2006) 146 LGERA 403* and *Roads and Traffic Authority (NSW) v McDonald* [2010] NSWCA 236).

In applying the before and after method, the valuer will:

- value the entire property before the compulsory acquisition of part of the land, using the relevant valuation method with full supporting evidence and
- value the retained portion of the property after the compulsory acquisition of part of the land, using the relevant valuation method with full supporting evidence,

with the difference between the valuations being the compensation amount.

Further, in the context of public purpose, Hemmings (1997) notes that, generally, in a 'before and after' valuation approach to the assessment of compensation, regard may not be had to the public purpose in the 'before' scenario but regard may be had to the public purpose in the 'after' scenario.

It should be noted that the difference may be positive or negative. If negative (where value after is less than value before), this is a source of compensation for the dispossessed. If positive (where value after is greater than value before – such as where smaller blocks of land sell at a higher rate than larger blocks of land), this is known as betterment and should not be double counted or should be deducted, if necessary, from the other value heads of compensation payable to the dispossessed.

2.5.2 Piecemeal method of valuation

The conventional application of the piecemeal method of valuation requires that the compensation for market value, injurious affection and severance should be separately calculated (if applicable) and then summed. As Rost and Collins (1990) succinctly note:

> As an alternative to the before and after method, a piecemeal assessment of compensation may be made. This comprises the market value of the land taken and appropriate amounts for severance, for injurious affection where this is incurred, and for all other consequential losses.
>
> (page 496)

Rost and Collins (1990) list the following factors to be considered in making a piecemeal assessment:

- impact on access;
- impact on water supply;
- impact on fencing;
- relative size of residue land remaining and
- effect on buildings (pages 496–497).

Jacobs (2015) notes that the piecemeal method:

> . . . requires the elements of the affected property to be broken into separate components whereafter a separate depreciation factor is applied to each component. Compensation may then be determined by adding together the amount allowed for the depreciation of each component.
>
> *(page 526)*

An alternative application of the piecemeal method of valuation may arise where the portion of land acquired is small and so may not be appropriately assessed using the before and after method, such as a narrow strip of land at the front of a block adjacent to a road that is acquired for road widening. The piecemeal method determines the value of the entire landholding, then derives an overall rate per square metre and applies this rate to the area of land acquired.

2.6 Summary and conclusions

Focusing on the key issues from a valuer's perspective, this chapter and Chapter 2 introduced the concepts and principles of statutory valuation, with statutory valuation for acquisition considered in Chapters 3, 4, 5 and 6 and statutory valuation for rating and taxing considered in Chapters 7 and 8. Regulatory valuation is then considered in Chapters 9 and 10.

Chapter 1 introduced the principles of value in statutory valuation, including the concepts of market value, liberal/conservative estimates of value, highest and best use, potentiality and confirmation of a foresight. This provided the essential foundation principles of statutory valuation that are then applied in the following chapters on statutory valuation.

This chapter sought to consider the practice of statutory valuation, including valuation methods, with an emphasis on the comparative method of valuation which has been repeatedly acknowledged by the Courts as the preferred method of valuation given its direct link to observable market transactions. The Court's decision in *Adams* was considered in detail, with an example illustrating how the sequential steps of accumulation, analysis, adjustment and application may lead to a transparent, explicit and logical assessment of value.

In the absence of comparable sales or where a check method of valuation may be required, alternative valuation methods may be accepted by the Courts depending on the specific circumstances of the matter under consideration. The Court's views on the income approach (including the capitalisation of income method, the related profits method and the discounted cash flow method) and the cost approach (including the replacement cost method, the related Australian summation method, the hypothetical development method and the related residual method) were considered, with further details and examples to be provided in Chapter 10. However, there are significant gaps between contemporary valuation theory and practice and the alternative valuation methods accepted by the Courts over the last century, such that care is required in the interpretation and application of historic case law.

While only of relevance to statutory valuation for compulsory acquisition purposes, the idiosyncratic before and after valuation method and piecemeal valuation method were considered in the context of injurious affection and severance, with examples to be provided in Chapter 4, concluding the review of valuation methods and conventions that may be preferred for application in statutory valuation.

Statutory valuation for acquisition is considered in Chapters 3, 4, 5 and 6. Chapter 3 introduces the principles of statutory valuation for acquisition, including the statutory framework and process, with a focus on the concept of disregard of the scheme in compensation assessment and the distinction between concepts of value and worth. This provides an insight into concepts of fairness and equity in statutory valuation in the context of the disregard of compulsion.

Chapter 4 explores the market value bases of compensation in acquisition for which statutory valuation may be required, including market value of the interest and the related market value compensation concepts of severance and injurious affection. This provides a broad understanding of the heads of compensation for acquisition on a value basis that may arise in statutory valuation.

Chapter 5 considers non-market value forms of compensation, including special value, which is noted to rarely arise, payment for solace arising from loss of home which is heavily dependent on the exact wording of the facilitating statute and the concept of disturbance for costs and losses suffered as a result of the acquisition. This provides a broad understanding of the heads of compensation for acquisition that are not necessarily on a value basis but that may arise in statutory valuation.

Chapter 6 concludes the chapters considering statutory valuation for acquisition by examining compensation for other interests such as easements and mortgages, mineral rights and profits à prendre, strata title and substrata, businesses and the emerging statutory valuation areas of native title and of indigenous cultural loss. This provides an appreciation of the diversity of interests that may be subject to acquisition and so may require valuation for the assessment of compensation.

Statutory valuation for rating and taxing is considered in Chapters 7 and 8. Chapter 7 introduces the principles of land as the basis for taxation through land tax, stamp duty and rates, considering the general principles of the valuation of land for taxation using the New South Wales statutory framework as an example. This provides a general understanding of the often arcane and artificial basis of statutory valuation upon which government collects very significant amounts of taxation revenue.

Chapter 8 explores the practice of statutory valuation for rating and taxing through exemptions, concessions and allowances that have evolved over the last century, together with contemporary issues such as mass appraisal, automated valuation models and artificial intelligence, the impact of major disasters on valuation for rating and taxing and the litigation process, including the valuer as an expert witness. This provides an insight into the continued evolution of statutory valuation to address issues as they arise over time, highlighting the importance of a principles-based approach to statutory valuation.

Regulatory valuation, considered in the context of the preparation of valuations for inclusion in financial reporting, is considered in Chapters 9 and 10. Chapter 9 explores the principles of regulatory valuation with regard to the inter-related roles of international accounting standards, international financial reporting standards and international valuation standards, focusing on the primacy of fair value in international financial reporting standards. This provides an understanding of the fundamental conceptual differences and bases between statutory valuation and regulatory valuation.

Finally, Chapter 10 concludes regulatory valuation by exploring the practice of valuation approaches and methods relevant for regulatory valuation such as the income approach and the cost approach and the various valuation methods there within. This provides an appreciation of the wider range of valuation approaches and methods available to the valuer for regulatory valuation, reflecting that, unlike statutory valuation, there is no requirement for defence in Court (negligence claims excepted).

References

Brown, D (2009) *Land acquisition*, Lexis Nexis Butterworths, Chatswood

Hemmings, N (1997) *Compensation, chapter in valuation principles and practice*, Australian Property Institute, Deakin.

Hyam, A (2014) *The law affecting valuation of land in Australia*, The Federation Press, Annandale

IVSC (2024) *International valuation standards*, IVSC, London

Jacobs, M (2015) *Law of compulsory land acquisition*, Thomson Reuters, Pyrmont

Parker, D (Ed) (2022) *Principles and practice of property valuation in Australia*, Routledge, Abingdon

Parker, D (2024) *Introduction to property valuation in Australia*, Routledge, Abingdon

Rawlinsons (2024) *Rawlinsons Australian cost handbook*, Rawlinsons, Perth

RICS (2024) *RICS valuation – global standards (red book)*, RICS, London

Rost, RO and Collins, HG (1990) *Land valuation and compensation in Australia*, 3rd Edition, Australian Institute of Valuers and Land Administrators, Deakin

Valuer General (2024) *Compensation following compulsory acquisition: guidance note*, Valuer General, Sydney

Wyatt, P (2023) *Property valuation*, Wiley Blackwell, Chichester

3 Compulsory acquisition

Concepts

3.1 Introduction

This book is an introduction to two very different valuation worlds in Australia – the complex and often arcane national realm of statutory valuation and the dynamic global empire of regulatory valuation. Statutory valuation is different from regulatory valuation, with the former being based on relevant applicable statutes and being capable of challenge on appeal in Court while the latter is based on relevant applicable regulations and incapable of challenge in Court (negligence claims excepted).

As an introductory textbook on statutory and regulatory valuation, this book is a companion to the Australian valuation textbooks, *Introduction to Property Valuation in Australia* (Parker, 2024) and *Principles and Practice of Property Valuation in Australia* (Parker, 2022), which provide a wider analysis of key principles underlying property valuation and current techniques and issues in the practice of property valuation for major sectors of the Australian property market.

Statutory valuation in Australia is most commonly encountered in valuations for rating, taxing and acquisition but may arise wherever a statute prescribes a basis of or approach to valuation. Regulatory valuation is most commonly encountered in valuations for financial statements but may arise wherever a regulation prescribes a basis of or approach to valuation.

Rating and taxing of land raise significant amounts of revenue for State Governments and Councils in Australia, based on statutory valuation. For example, the NSW 2025–26 Budget Statement estimates land tax revenue for 2024–2025 to be $8.2 billion with the City of Brisbane Annual Plan and Budget 2024–2025 estimating proposed rates and utility charges to be $1.51 billion.

Acquisition of land, property and businesses commonly arises for major infrastructure projects and may be by agreement or by compulsion, based on statutory valuation. For example, such major infrastructure projects in Australia currently include the multi-billion-dollar Sydney Metro, Melbourne Suburban Rail Loop and Melbourne to Brisbane Inland Rail projects.

Valuation of land and property holdings based on regulatory valuation arises for all reporting entities in Australia that publish financial statements upon which it is reasonable to expect that users will rely for decision making. For example, at June 2025 there are over 2,000 reporting entities listed on the Australian Stock Exchange which each publish financial statements, with the financial statements for Goodman Group, as at 30 June 2024, showing the fair value of total property assets to be $20.2 billion, based on regulatory valuation.

Focusing on the key issues from a valuer's perspective, the concepts and principles of statutory valuation are introduced in Chapters 1 and 2, with statutory valuation for acquisition considered in Chapters 3, 4, 5 and 6 and statutory valuation for rating and taxing considered in Chapters 7 and 8. Regulatory valuation is then considered in Chapters 9 and 10.

DOI: 10.1201/9781032616957-3

Chapter 1 introduced the principles of value in statutory valuation, including the concepts of market value, liberal/conservative estimates of value, highest and best use, potentiality and confirmation of a foresight. This provided the essential foundation principles of statutory valuation that are then applied in the following chapters on statutory valuation.

Chapter 2 considered the practice of statutory valuation including valuation methods for statutory valuation with an emphasis on the comparative method of valuation, together with statutory valuation conventions such as the before and after method and the piecemeal method. This provided an understanding of the valuation methods and conventions that may be preferred in the following chapters on statutory valuation.

Statutory valuation for land acquisition is considered in Chapters 3, 4, 5 and 6, supported by an evolving case study in Appendix 1 of each chapter, including the principal bases of compensation being:

- compensation for market value for the interest acquired;
- compensation for severance;
- compensation for injurious affection;
- compensation for special value;
- compensation for solatium or disadvantage resulting from relocation and
- compensation for disturbance.

This chapter introduces the principles of statutory valuation for land acquisition, including the statutory framework and process, with a focus on the concept of the disregard of the scheme in compensation assessment and the distinction between concepts of value and worth. This provides an insight into concepts of fairness and equity in statutory valuation in the context of the requirement to disregard the compulsory nature of the acquisition.

Chapter 4 explores the market value bases of compensation in acquisition for which statutory valuation may be required, including market value of the interest and the related market value compensation concepts of severance and injurious affection. This provides a broad understanding of the heads of compensation for acquisition on a value basis that may arise in statutory valuation.

Chapter 5 considers non-market value forms of compensation including special value, which is noted to rarely arise, payment for solace arising from loss of the primary place of residence and the concept of disturbance for costs and losses suffered as a result of the acquisition. This provides a broad understanding of the heads of compensation for acquisition that are not necessarily on a value basis but which may arise in statutory valuation.

Chapter 6 concludes the chapters considering statutory valuation for acquisition by examining compensation for other interests such as easements and mortgages, mineral rights and profits à prendre, strata title and substrata, businesses and the emerging statutory valuation areas of native title and indigenous cultural loss. This provides an appreciation of the diversity of interests that may be subject to acquisition and so may require valuation for the assessment of compensation.

Statutory valuation for rating and taxing is considered in Chapters 7 and 8. Chapter 7 introduces the principles of land as the basis for taxation through land tax, stamp duty and rates, considering the general principles of the valuation of land for taxation using the New South Wales statutory framework as an example. This provides a general understanding of the often arcane and artificial basis of statutory valuation upon which Government collects very significant amounts of taxation revenue.

Chapter 8 explores the practice of statutory valuation for rating and taxing through exemptions, concessions and allowances that have evolved over the last century, together with contemporary issues such as mass appraisal, automated valuation models and artificial intelligence, the impact of major disasters on valuation for rating and taxing and the litigation process including the valuer as an expert witness. This provides an insight into the continued evolution of statutory valuation to address issues as they arise over time, highlighting the importance of a principles-based approach to statutory valuation.

Regulatory valuation, considered in the context of the preparation of valuations for inclusion in financial reporting, is considered in Chapters 9 and 10. Chapter 9 explores the principles of regulatory valuation with regard to the inter-related roles of international accounting standards, international financial reporting standards and international valuation standards, focusing on the primacy of fair value in international financial reporting standards. This provides an understanding of the fundamental conceptual differences and bases between statutory valuation and regulatory valuation.

Finally, Chapter 10 concludes regulatory valuation by exploring the practice of valuation approaches and methods relevant for regulatory valuation such as the income approach and the cost approach and the various valuation methods there within. This provides an appreciation of the wider range of valuation approaches and methods available to the valuer for regulatory valuation reflecting that, unlike statutory valuation, there is no requirement for defence in Court (negligence claims excepted).

Accordingly, this chapter will now consider:

- enabling and facilitating statutes;
- fairness, just terms and hardship;
- interests in land and
- public purpose and disregard of the public purpose.

3.2 Enabling and facilitating statutes

This section seeks to guide the valuer through the valuation aspects of the key issues concerning facilitating and enabling statutes from a valuer's viewpoint. Greater detail on the legal aspects of the key issues may be found in Jacobs (2015), Hyam (2014), Brown (2009) and Newton and Conolly (2017).

3.2.1 Acquisition under statutory powers

Each of the jurisdictions of the Commonwealth, the six States and two Territories has legislation enabling agencies at each level of government (including councils) to compulsorily acquire land for specified public purposes. As Brown (2009) notes:

> The citizens of the nation accept that a landowner who is deprived of the land may be unfortunate or unlucky but the taking of their land, for the good of the community as a whole, is necessary. It may be regarded by the community as a regrettable but necessary episode. The legislation aims to ensure that the owner does not suffer a loss from a financial point of view.
>
> (page 1)

In Australia, the nouns *compulsory acquisition* are commonly referred to as *resumption*, with the verb *to compulsorily acquire*, being commonly referred to as *to resume*. Further, compulsory acquisition is referred to as *expropriation* in Canada and South Africa and as *eminent domain* in the USA.

3.2.1.1 Acquisition by agreement or compulsion

Land may be acquired by government either by agreement or through compulsory acquisition. While information concerning the proportion of land acquired by agreement or by compulsion is published in some States as annual activity totals, the data is not synchronised, so it is not possible to determine, for example, exactly what proportion of Valuer General determinations following compulsory acquisition in New South Wales are appealed to the Land and Environment Court.

Parker (2024) notes that, in New South Wales, it is possible to conclude that typically 76%–88% of acquisitions are settled by agreement, leaving typically 12%–24% to be acquired compulsorily. However, in New South Wales, the dispossessed and the acquiring authority are required to mediate in the Land and Environment Court prior to the matter proceeding to a hearing before a Judge or Commissioner. Of those matters that are appealed to the Land and Environment Court, typically 62%–86% are finalised by agreement, leaving only 14%–38% (or 10–17 matters) requiring a hearing.

Effectively, therefore, a very high proportion of compulsory acquisitions in New South Wales are settled by agreement (possibly around 97%–98%), with only around 2%–3% where the parties are unable to agree and a decision is made by a Judge or Commissioner of the Land and Environment Court (Parker, 2024).

3.2.1.2 Enabling and facilitating statutes

Compulsory acquisition in Australia is commonly a function of the interaction of two statutes:

- an enabling statute and
- a facilitating statute.

An enabling statute will make provision for a government entity to compulsorily acquire land for the purposes set out in the statute. Enabling statutes are generally used in New South Wales, Victoria and South Australia, while the Commonwealth, Queensland, Western Australia and Tasmania generally combine enabling and facilitating statutes (though this does not preclude their use of enabling statutes). Brown (2009) notes that, in the Northern Territory and the Australian Capital Territory, the legislation enables compulsory acquisition to be carried out for purposes which are within the scope of a Territory's power to make laws.

By way of example, enabling statutes such as Water Acts will usually make provision for the compulsory acquisition of land and property for the purposes of water infrastructure, and Railways Acts will usually make provision for the compulsory acquisition of land and property for the purposes of rail infrastructure.

For example, the *NSW Education Act 1990* states:

> 125 Acquisition and disposal of land
>
> (1) The Minister may, for the purposes of this Act or jointly for those purposes and purposes of or associated with public education or recreation, acquire land (including an interest in land) by agreement or by compulsory process in accordance with the Land Acquisition (Just Terms Compensation) Act 1991.

The valuer should check that the enabling statute, which is usually referred to in the acquisition notice, does actually enable the acquisition of land for the public purpose specified in the acquisition notice. This may be clear, such as where a Railways Act enables the acquisition of land for the purposes of rail infrastructure and the public purpose is the construction of a railway line. It may be less clear, however, where the public purpose is, for example, the construction of a car park, and the valuer may need to seek legal advice from the client.

Generally, the enabling statute will be linked to a facilitating statute which provides the machinery that is to be followed for the compulsory acquisition process, with the relevant facilitating statutes being:

- Commonwealth – *Lands Acquisition Act* 1989
- New South Wales – *Land Acquisition (Just Terms Compensation) Act* 1991
- Queensland – *Acquisition of Land Act* 1967
- Victoria – *Land Acquisition and Compensation Act* 1986
- South Australia – *Land Acquisition Act* 1969
- Western Australia – *Land Administration Act* 1997
- Tasmania – *Land Acquisition Act* 1993
- Northern Territory – *Lands Acquisition Act* 1978
- Australian Capital Territory – *Lands Acquisition Act* 1994

While there are some similarities, there are also significant differences between the facilitating statutes in each jurisdiction, and the valuer should have careful regard to the specific facilitating statute in the jurisdiction within which a valuation is being undertaken.

The valuer should check the facilitating statute, which will generally provide not only the bases for compensation (with only such bases specified being available to the valuer to ground a payment of compensation) but also any time frames which need to be adhered to.

3.2.1.3 An interest in land

Only those who have an 'interest in land' are generally entitled to compensation. Some interests are clear and very secure, including the registered proprietor, registered lessees and owners of the benefit of an easement. Others are less secure, such as unregistered lessees and those who occupy land under verbal agreements or other understandings. This often arises with businesses where the land is owned by the directors of the business, but the business entity doesn't have a formal lease.

Recently, the courts have emphasised that only those interests which are enforceable and transferable are entitled to receive compensation when that interest is extinguished by a compulsory acquisition. This has meant that occupiers under oral leases or licences are generally not entitled to any compensation at all.

Interests in land are considered further in Section 3.4.

3.2.2 Generic model – the acquisition process

The following provides a generic model for the acquisition process as a series of steps. It should be noted that not every step may be provided in every State or Territory's facilitating statute, and the valuer should check the relevant facilitating statute carefully.

3.2.2.1 *Preliminary negotiation*

Following an announcement by government of a proposed infrastructure project, with associated media attention, landowners may suspect that their land may be acquired and will often be expecting to hear from the relevant authority.

The relevant authority may either write to the landowner or meet with the landowner to advise of the proposed acquisition and provide an indicative time frame for the various steps in the process.

The relevant authority, which will be the acquiring authority, may obtain a preliminary valuation, and the landowner may or may not choose to obtain a preliminary valuation with which to commence negotiations.

A period of negotiation will then occur between the landowner and the acquiring authority where the parties attempt to agree the compensation to be paid – in a high proportion of cases, the compensation payable is agreed by negotiation between the parties.

Facilitating statutes generally provide for acquisition by agreement consistent with the framework that would apply if the acquisition were to be compulsory. While this may be considered to be a period of voluntary bargaining, the landowner will be aware of the acquiring authority's compulsory acquisition powers that will loom over the negotiation process, often referred to as negotiation under the cloud of compulsion or in the shadow of compulsion.

If the acquiring authority and the landowner are able to agree on compensation, the acquisition may be settled. If not, the acquiring authority will proceed with the process for compulsory acquisition.

3.2.2.2 *Formal notice of acquisition*

If the acquisition is to be by compulsion, the acquiring authority will provide the landowner with a formal notice advising of the proposed acquisition, the reason for the acquisition (generally referred to as the *public purpose*) and the relevant enabling statute.

The acquiring authority may be required to publish the formal notice in a government publication (such as the *Government Gazette*). Hemmings (1997) notes that:

- upon publication of the notice, all specified interests of specified persons in the land are thereupon vested in the acquiring authority and
- the acquiring authority is vested with an estate in fee simple, discharged and freed from all trusts, obligations and interests, contracts, licences, rates, charges, easements and all other restrictions and encumbrances.

At the point of publication of the formal notice, ownership of the land and property transfers to the acquiring authority and the landowner's interest converts to a claim for compensation, with the date of publication usually being the date of valuation. Parker (2019) notes that acquisition and dispossession are instant and simultaneous (though some acquiring authorities may allow continued occupation), while compensation may be agreed at a later time, although a proportion may be payable in advance. Effectively, the landowner cannot hold up the project for which the land is being acquired, as the concept of acquisition is severed from the concept of compensation.

The date of the publication of the acquisition notice is usually the date of valuation and the date from which various time frames specified in the facilitating statute will usually be determined.

3.2.2.3 Formal claim for compensation by the landowner

The landowner will make a formal claim for compensation to the acquiring authority. This claim is usually supported by relevant evidence such as the report of an independent valuer instructed by the landowner and other expert reports, if required, such as from a planner, hydrologist, heritage expert and so forth.

3.2.2.4 Period of negotiation

The acquiring authority and the dispossessed landowner may continue to negotiate the compensation payable with the aim of reaching agreement without the need for litigation.

In the event that the compensation payable cannot be agreed, the acquiring authority may refer the matter to an independent party who will determine the compensation payable.

3.2.2.5 Independent determination

The independent determination of compensation is then usually made by the Valuer General of the relevant State or Territory, often after considering a valuation and submission from each of the acquiring authority and the dispossessed landowner.

The independent party, usually the Valuer General, will then issue a determination of compensation specifying the amount of compensation that the acquiring authority will pay to the dispossessed landowner.

3.2.2.6 Payment of compensation

The acquiring authority then pays the amount of compensation determined by the independent party, usually the Valuer General, to the landowner, often with interest.

3.2.2.7 Judicial determination

If the dispossessed landowner is not satisfied with the amount of compensation determined by the independent party, usually the Valuer General, they may lodge an objection with the relevant Court or Tribunal.

A judicial officer in the relevant Court or Tribunal will then hear evidence from the acquiring authority and the dispossessed landowner and will make a determination of compensation.

Such determination is usually final and may usually only be appealed on the basis of an error of law.

3.2.3 Case study – the acquisition process

Brown (2009) considers the steps in the acquisition process for each State facilitating statute at length. As an example of the steps in the acquisition process within a facilitating statute, the *NSW Land Acquisition (Just Terms Compensation) Act 1991* will be considered further below.

The objects of the *NSW Land Acquisition (Just Terms Compensation) Act 1991* include compensation on just terms at not less than the market value of the land at the date of acquisition:

3 Objects of Act

(1) The objects of this Act are–

(a) to guarantee that, when land affected by a proposal for acquisition by an authority of the State is eventually acquired, the amount of compensation will be not less than the market value of the land (unaffected by the proposal) at the date of acquisition, and
(b) to ensure compensation on just terms for the owners of land that is acquired by an authority of the State when the land is not available for public sale, and
(c) to establish new procedures for the compulsory acquisition of land by authorities of the State to simplify and expedite the acquisition process, and
(d) to require an authority of the State to acquire land designated for acquisition for a public purpose where hardship is demonstrated, and
(e) to encourage the acquisition of land by agreement instead of compulsory process.

(2) Nothing in this section gives rise to, or can be taken into account in, any civil cause of action.

3.2.3.1 Preliminary negotiation

Significantly, the objects of the Act include encouragement of acquisition by agreement. Section 10A(2) requires the acquiring authority to make a *genuine attempt* to acquire the land by agreement for at least six months before giving a proposed acquisition notice, though this may be shortened by agreement or by the Minister in certain circumstances. If the acquiring authority and the landowner reach agreement, they can transfer the land by a contract of sale (or a grant of lease or transfer granting easement), or they can agree that the land be acquired by compulsory process under s30 (and s29 for public authorities). Negotiations may continue even after the service of the proposed acquisition notice.

3.2.3.2 Formal notice of acquisition

Section 11 requires the acquiring authority to give written notice of the compulsory acquisition (being a proposed acquisition notice or PAN) to all owners of interests in the land (s12) at least 90 days before the land is acquired, though this may be shortened by agreement or by the Minister in certain circumstances (s13).

After giving the notice, the acquiring authority must advise the Registrar General (s17) and the Valuer General (s18) and publish the notice in the *Government Gazette* (s19). Section 20 provides:

(1) On the date of publication in the Gazette of an acquisition notice, the land described in the notice is, by force of this Act –

(a) vested in the authority of the State acquiring the land, and
(b) freed and discharged from all estates, interests, trusts, restrictions, dedications, reservations, easements, rights, charges, rates and contracts in, over or in connection with the land.

Therefore, at the date of gazettal, all the acquired interests in the land transfer to the acquiring authority and are replaced, for the dispossessed, by the right to receive compensation. The project for which the land is being acquired may, therefore, proceed before the amount of compensation has been agreed, with the dispossessed unable to hold up the project.

Generally, landowners cannot force government agencies to acquire their land even if it is zoned for public purposes (such as open space, drainage and protection of native vegetation). The exception is where the landowner can demonstrate that they are suffering 'hardship'. This is defined as a set of circumstances where the owner can't sell their land and they need funds for family, social or other economic reasons. If a landowner can establish hardship, they are entitled to market value of the land. They may receive, but are not entitled to receive, additional compensation for special value of land, severance, disturbance and disadvantage resulting from relocation. It is a discretionary matter for the acquiring authority whether the additional compensation should be paid. Hardship is considered further in Section 3.3.2.

3.2.3.3 Formal claim for compensation by the landowner

Section 15 of the Act requires that the proposed acquisition notice request the landowner to lodge a claim for compensation within 60 days of the proposed acquisition notice (s15). Section 37 provides an entitlement to compensation and s39 requires the dispossessed to lodge a claim for compensation.

Section 41(1) requires the acquiring authority, within seven days after it compulsorily acquires the land, to provide the Valuer General with a list of the issues that the acquiring authority believes are relevant to the determination of the amount of compensation by the Valuer General.

3.2.3.4 Period of negotiation

While the Act provides a pathway from the issue of the proposed acquisition notice to a determination by the Valuer General over a specified period of time, the acquiring authority and the dispossessed landowner may continue to negotiate compensation during this period.

3.2.3.5 Independent determination

With the parties having provided their opinions of compensation under s39 and s41(1), s47 provides for the Valuer General to determine the compensation payable within 45 days (s42(1)), with the process outlined in s41.

3.2.3.6 Payment of compensation

Section 44(2) provides that payment of the compensation is to be made within 28 days of the receipt by the acquiring authority of a claim for compensation, deed of release and indemnity (duly completed) and any relevant documents of title.

The offer of compensation is deemed to be accepted 90 days after the compensation notice is provided if the dispossessed landowner has either not formally accepted or lodged an objection with the Land and Environment Court.

Section 48 makes provision for advanced payments of compensation and s49 makes provision for payment of interest on compensation.

3.2.3.7 Judicial determination

The dispossessed owner may appeal the Valuer General's determination of compensation by lodging an objection with the Land and Environment Court within 90 days after receiving a compensation notice (s66(1)) and the Land and Environment Court shall hear and dispose of the claim (s66(2)). The Court has the ability to hear objections lodged outside 90 days if satisfied that there is good cause for the person's failure to lodge an objection within the specified period (s66(3)).

3.2.4 ANZVGP 113 Valuations for compensation and compulsory acquisition

The Australian Property Institute has published *ANZVGP 113 Valuations for Compensation and Compulsory Acquisition* (API, 2021a) and *ANZVGP 113.1 Resource Pack* (2021b). As at 15 January 2025, when each was accessed online, neither publication had been withdrawn.

ANZVGP 113 Valuations for Compensation and Compulsory Acquisition states that the objective of a Guidance Paper is to clarify best practices and that a Court or Tribunal may consider the contents in deciding whether an API member acted to a standard required by law. *ANZVGP 113.1 Resource Pack* states that Guidance Papers and Resource Packs are a guide and measure of acceptable professional practice.

ANZVGP 113 Valuations for Compensation and Compulsory Acquisition provides useful information definitions (2.0), guidance on instructions and fees (3.0), basis of compensation assessment (4.0), heads of compensation (5.0), inspection and enquiry (6.0), report content (7.0) and post assessment consultancy (8.0).

Importantly, *ANZVGP 113 Valuations for Compensation and Compulsory Acquisition* addresses ethical obligations (3.2), focusing on valuers only operating within the limits of their qualifications and experience and assessing potential conflicts of interest. Further, the important distinction between acting as an expert (providing independent advice regardless of the client) and as an advocate (conducting negotiations on the client's behalf) are addressed in Sections 3.3.1 and 3.3.2, including the fundamentally important provision that an API member cannot act as an expert and as an advocate in the same matter.

Care is, however, required regarding compliance with and departures from International Valuation Standards (IVS) (1.2) and the use of IVS definitions in value of land taken (5.1) in the context of relevant facilitating statutes and case law.

3.2.5 Example – enabling and facilitating statutes

Appendix 1 contains a case study which will be used to illustrate the various aspects of compulsory acquisition considered in the chapter.

The proposed acquisition notice stated that the land was to be acquired *for the purposes of the Roads Act . . . in connection with the construction, operation and maintenance of the Happytown Freeway.*

Concerning the enabling statute, the proposed acquisition notice refers to the *Roads Act 1993* (NSW) which includes the following land acquisition power:

177 Power to acquire land generally

(1) The Minister, TfNSW or a council may acquire land for any of the purposes of this Act.
(3) Without limiting subsection (1), TfNSW may also acquire land that it proposes to declare to be TfNSW development land.

Further, Section 11 of the *Roads Act* 1993 provides:

11 Land acquired by Minister

(1) The Minister may, by notice published in the Gazette, dedicate any land acquired by the Minister under Division 1 or 2 of Part 12 as a public road.

Section 48 of the *Roads Act* 1993 provides that:

48 Freeways

(1) The Minister may, by order published in the Gazette, declare to be a freeway any main road that is designed to facilitate the movement of motor traffic.
(2) On the publication in the Gazette of an order declaring a main road to be a freeway, TfNSW becomes the owner of the land on which the freeway is situated.

and s71 provides that:

71 Powers of roads authority with respect to road work

A roads authority may carry out road work on any public road for which it is the roads authority and on any other land under its control.

Accordingly, the *Roads Act 1993* enables the acquisition of land for the construction, operation and maintenance of a freeway.

Concerning the facilitating statute, as Happytown is located in country New South Wales, the relevant facilitating statute is the *Land Acquisition (Just Terms Compensation) Act 1991*.

Assume that preliminary negotiations have been underway between the acquiring authority and Mickey, Minnie, Goofy, Donald and Pluto for over six months, with the acquiring authority making a genuine attempt to acquire the land by agreement. As no agreement could be reached, the acquiring authority has issued a proposed acquisition notice to Mickey, Minnie, Goofy, Donald and Pluto on and dated 1 July 2024, with an acquisition date of 29 September 2024. The acquiring authority advised the Registrar General and the Valuer General and published the proposed acquisition notice in the NSW *Government Gazette* on 1 July 2024.

3.3 Fairness – just terms and hardship

This section seeks to guide the valuer through the valuation aspects of fairness through the lenses of just terms and hardship. Greater detail on the legal aspects of the key issues may be found in Jacobs (2015), Hyam (2014), Brown (2009) and Newton and Conolly (2017).

3.3.1 Fairness – just terms

The Australian Constitution provides the Commonwealth with the power to acquire property *on just terms from any state or person for any purpose in respect of which the parliament has the power to make laws* (s51), such as the acquisition of land for defence purposes on just terms. It should be noted that such acquisition may be by agreement or compulsorily.

State Governments may enact legislation to compulsorily acquire land either with or without compensation. However, as the community generally expects State Governments to protect private land rights, in each State and Territory the legislation requires the payment of compensation based on the market value of the land being acquired as well as other heads of compensation.

Further, the Federal, State and Territory Courts deal with disputes between acquiring authorities and dispossessed parties, with their decisions providing a secondary source of compulsory acquisition law. Readers will note extensive references through the statutory valuation sections of this book to various Court decisions to which valuers must have regard when undertaking statutory valuations.

Hemmings (1997) notes that compensation for compulsory acquisition is not part of the common law but is solely a matter of statutory entitlement in Australia and the assessment of compensation is limited by that legislation.

For example, in NSW Section 54(1) of the *Land Acquisition (Just Terms Compensation) Act 1991* provides for:

54 Entitlement to just compensation

1 The amount of compensation to which a person is entitled under this Part is such amount as, having regard to all relevant matters under this Part, will justly compensate the person for the acquisition of the land.

Therefore, in Australia compulsory acquisition is generally based on the principle of *on just terms*, with the market value of land taken and other forms of compensation, as permitted by the various States' and Territories' facilitating statutes (statutory entitlement), being assumed to provide just compensation. There is, however, an underlying tension that the payment of statutory entitlement may not result in compensation that is just.

Such tensions are not confined to Australia, with Du Plessis (2018) examining the tensions between expropriation statutes in South Africa and the South African Constitution, where the Constitution requires just and equitable compensation even if this compensation is less than market value:

> Therefore, when calculating compensation the courts should first determine the market value of the property (since it is easily quantifiable), and then, based on the list in Section 25(3) (in the constitution), adjust the compensation amount either upwards or downwards. Market value is to be elevated to a central or starting position in determining compensation, with the factors in the discretion of the judge determining by how much the compensation should be increased or decreased.
>
> (page 195)

Effectively, Australian acquisition legislation seeks to provide the dispossessed owner with an amount that will justly compensate, being the full money equivalent of the property acquired by compulsion.

However, therein lies the tension for valuers. While statutory entitlement may seek to provide compensation that is just, if the facilitating statute does not have a provision permitting the payment of compensation for an item claimed, then there is no basis upon which that item can be compensated by the acquiring authority. While the dispossessed landowner, in such circumstance, is unlikely to consider the compensation to be just, it is a product of balancing the interests of the dispossessed and the wider community/taxpayer consistent with the balance that underlies *Spencer*, as considered in Section 1.2.2.

A common example of apparent unfairness or unjust terms arises from the compulsory acquisition of the primary place of residence from an elderly widow. The widow may have lived in the house for 70 years with her late husband and raised her family in the house, such that to her (to quote from the classic movie *The Castle*) 'It's not a house. It's a home'. Accordingly, for the compensation to be perceived as fair and the acquisition to be on just terms, the widow may be likely to be seeking compensation for the worth of her home, but the enabling statute may not permit compensation for worth of a home, instead providing compensation for the value of a house.

3.3.1.1 Pre-facilitating statutes period

Prior to the introduction of the various facilitating statutes, the Courts adopted the concept of Judge determined compensation for *value to the owner*, which was a combination of market value, special value, severance and disturbance.

This was because the earlier statutes did not set out the relevant heads of compensation, such that the Courts had to develop the applicable compensation principles, such as those contained in *Spencer v Commonwealth* (1907) 5 CLR 418. Special value, in this context, may be payable where financial advantage to the owner above and beyond market value could be demonstrated.

It may be contended that the ability of a Judge to award compensation for special value contributed to the sense of fairness in compensation, with acquisition then being considered to be on just terms.

3.3.1.2 Post-facilitating statutes period

Following the introduction of the various facilitating statutes, that which may be compensated was codified in the respective statutes, and compensation must now be assessed strictly in accordance with the facilitating statute.

For example, in New South Wales, the facilitating statute is the *Land Acquisition (Just Terms Compensation) Act 1991*, with s55(a) and s56 providing for the payment of market value, s55(e) and s60 providing for the payment of solatium and s55(b) and s57 providing for the payment of special value, which are defined as:

Solatium:

Section 55(e):

In determining the amount of compensation to which a person in entitled, regard must be had to the following matters only (as assessed in accordance with this Division):

(e) the disadvantage resulting from relocation,

Section 60:

(1) In this Act:

"disadvantage resulting from relocation" means non-financial disadvantage resulting from the necessity of the person entitled to compensation to relocate the person's principal place of residence as a result of the acquisition.

Special value

Section 55(b):

In determining the amount of compensation to which a person in entitled, regard must be had to the following matters only (as assessed in accordance with this Division):

(b)any special value of the land to the person on the date of its acquisition,

Section 57:

In this Act:

> "special value" of land means the financial value of any advantage, in addition to market value, to the person entitled to compensation which is incidental to the person's use of the land.

Therefore, a Judge may award market value for the loss of the house and an amount for solatium to the elderly widow to reflect the disadvantage resulting from relocation but is unlikely to be able to award compensation for special value to reflect the worth of her home to her.

Accordingly, the facilitating statutes may constrain the ability of a Judge to award compensation for worth, which contributes to the sense of unfairness in compensation, with acquisition then being considered not to be on just terms.

3.3.2 Fairness – hardship

Where the government places a notation on title or changes the planning controls to indicate that land is intended to be acquired for public purposes, the relevant land value can be adversely impacted. Where an owner of such land cannot sell their land and pressing personal, domestic or social reasons arise, the owner can request the acquiring authority to acquire their land. This is called hardship.

For example, some agencies or councils may not have the budget to acquire land immediately. However, the notation on title (created by the registration by the acquiring authority of a plan of acquisition) or the zoning or reservation under the planning controls means that the land will be acquired at some point in the future for development for a public purpose.

There may be a long time period between the notation on title or the land being designated under the planning controls and the acquisition occurring, with such delay potentially causing hardship to the landowner because they either cannot sell their land or cannot sell it for its market value.

Two States provide for a landowner to compel an authority to compulsorily acquire their land, being New South Wales (s21–28, *Land Acquisition (Just Terms Compensation) Act 1997*) and Victoria (s76, *Land Acquisition and Compensation Act 1986*) (Brown, 2009). Such a request for acquisition usually requires a supporting reason (such as an inability to sell due to the impending acquisition and pressing personal, domestic or social reasons) and may limit the heads of compensation claimable. For example, owners may claim, but are not entitled to, other heads of compensation such as disturbance.

3.4 Interests in land

This section seeks to guide the valuer through the valuation aspects of the key issues concerning interests in land from a valuer's viewpoint. Greater detail on the legal aspects of the key issues may be found in Jacobs (2015), Hyam (2014), Brown (2009) and Newton and Conolly (2017).

Brown (2009, page 24) and Jacobs (2015, page 41) identify and list the definitions of an interest in land in the various Commonwealth, State and Territory facilitating statutes, with Brown (2009) concluding that *The term 'interest' defies exact definition* (page 25) and that *There is no closed, precise list of what is or is not an interest in resumed land* (page 27).

It is important that the valuer identify every compensable interest in the land to be acquired. As noted earlier, not all interests are compensable interests. Only those interests which are enforceable and alienable (which can be bought and sold or transferred) are compensable interests. Parker (2019) notes that land ownership in Australia stretches from the centre of the earth to the outermost part of the heavens (*usque ad coelum et ad inferos*), subject to the usual restrictions (e.g. underground minerals, surface level swans, airspace for flights, etc).

Therefore, if a parcel of land is acquired, this includes all the relevant subterranean interests as well as the interests on the ground such as buildings and interests above the ground such as aerial billboards, vertical development rights and so forth. Valuers must search the title and all relevant dealings on title and undertake a physical inspection of the acquired land to identify and confirm interests in land.

3.4.1 Identification of interests

When undertaking a valuation for compensation arising from compulsory acquisition, it is important to determine the nature and extent of the interests being valued. The acquisition of an investment property may require compensation to be paid to a range of parties including:

- the owner of the fee simple estate (the registered proprietor);
- owners of lots in a strata plan or community title scheme;
- the owners corporation in a strata plan or the manager of the community scheme;
- leaseholders including:
    - registered lessees;
    - unregistered lessees;
    - lessees who are holding over their lease or who have a tenancy at will;
- mortgagees;
- owners of the benefit of easements and
- those who are entitled to trees or soil from land (called a profit a prendre, though rare in Australia). (Hyam, 2014, pages 27–41).

Other interests in land can include purchasers under an uncompleted contract for sale or option agreement.

Recent cases have illustrated that licensees or those with very insecure or short term interests that are not enforceable and are not transferable are unlikely to have any compensable interest in the land (see *Olde English Tiles Australia Pty Ltd v Transport for New South Wales [2022] NSWCA 108*; *Dial A Dump Industries Pty Ltd v Roads and Maritime Services* [2017] Court of Appeal 73).

It is important to establish the exact nature of the interest for which compensation is being determined, as the compensation payable for a long-term tenancy may differ significantly from the compensation payable for a short term tenancy.

Freehold, leasehold and licence interests may also include a business as a going concern for which compensation will be required to be determined, such as a newsagent or snack bar operating in a shop on the ground floor of an office or residential tower that is being compulsorily acquired. Where multiple interests exist within one property that is being acquired, it is important to ensure that there is no double counting.

Table 3.1 Case study – parties and interests in land

Party	*Freehold*	*Leasehold*	*Lic/TcyWill*	*Rem/Rev*	*Mortgage*	*Easement*	*Business*
Mickey	√						√
Minnie		√					
Donald			√				√
Daisy		√					
Pluto		√					√
Goofy			√				

Source: Author

3.4.2 Example – identification of interests

Appendix 1 contains a case study which will be used to illustrate the various aspects of compulsory acquisition considered in Chapters 3, 4, 5 and 6.

Concerning the identification of interests, it is important that the valuer identify each of the interests in land held by each of the parties subject to compulsory acquisition. For completeness, it may be helpful to construct a matrix identifying the range of parties and the range of interests in land from the case study, as shown in Table 3.1.

Within the leasehold interests, the following forms of leasehold interest may be identified:

- Minnie – lease;
- Daisy – periodic tenancy and
- Pluto – lease.

Having identified the parties and interests in land held, the valuer is now able to consider the different forms of compensation to which each party may be entitled and which will be considered in Chapters 4, 5 and 6.

3.5 Public purpose and disregard of the public purpose

This section seeks to guide the valuer through the valuation aspects of the key issues concerning public purpose and the disregard of public purpose from a valuer's viewpoint. Greater detail on the legal aspects of the key issues may be found in Jacobs (2015), Hyam (2014), Brown (2009) and Newton and Conolly (2017).

The public purpose is the works for which the land and property are being acquired. In the example of a high school, the public purpose may be described simply as *for the development of XYZ High School*. As noted in Section 3.2.1.2, the valuer should check that the enabling statute, which is usually referred to in the acquisition notice, does actually enable the acquisition of land for the public purpose specified in the acquisition notice.

This may be clear, such as where a Railways Act enables the acquisition of land for the purposes of rail infrastructure and the public purpose is the construction of a railway line. It may be less clear, however, if the public purpose were to be, for example, the construction of a car park and the valuer may need to seek legal advice from the client.

The public purpose is important because it is generally to be disregarded in determining the market value of land and property that is compulsorily acquired, commonly known as the public policy disregard. For example, the extension of a train line and construction of a station may be expected to positively influence surrounding residential land values. However, in the assessment of compensation for surrounding residential land and property, such positive influence is to be disregarded in the valuation and the value determined as if the train line and train station did not exist.

The public policy intention of the disregard of the public purpose is to prevent landowners benefitting from the impact of public works while providing them with a fair payment of compensation.

3.5.1 Public purpose

Brown (2009) identifies and lists the relevant provisions concerning public purpose in the Commonwealth, State and Territory facilitating statutes (pages 53–55), with Jacobs (2015) noting 'There must be a causal link between the stated purpose of the resumption and the "purposes of the Act", failing which the resumption will be beyond the ambit of the statutory power' (page 99).

Concerning the principle of public purpose, in *Clunies-Ross v Commonwealth of Australia* (1984) 55 ALR 609, the Court considered the public purpose for land acquisition in the Cocos (Keeling) Islands. The owner of most of the land (*'king' of the Cocos Islands*) had previously sold the land to the Commonwealth but retained ownership of a large house and gardens. The Commonwealth later sought to compulsorily acquire the house and gardens but gave no specific reason for the acquisition.

When challenged, the Commonwealth stated that the land was required for the political, social and economic advancement of the people of the islands – whereas, in reality, the Commonwealth simply wanted to get the landowner off the islands permanently and had no particular need for the land. The Court held that the power of acquisition did not extend to the taking of the land for the sole purpose of depriving the owner of it, giving rise to the principle that land may only be compulsorily acquired for a lawful purpose.

As a general principle, acquiring authorities should only use compulsory acquisition powers for the public purpose. Therefore, as a general principle, if the public purpose is the creation of a highway, compulsory acquisition powers should only be used to acquire that amount of land required to create the highway.

Further, as a general principle, acquiring authorities should only acquire that land which is required for the public purpose and not acquire more land than is required with the intention of on-selling it.

In *Mac's Pty Ltd v Minister Administering Local Government Act 1993 (NSW)* (2007) 155 LGERA 362, the Court rejected the compulsory purchase of land for resale as a public purpose. While this decision was set aside by the Court of Appeal in *Parramatta City Council v R&R Fazzolari Pty Ltd* (2008) 162 LGERA 1, it was largely reinstated on appeal to the High Court in *R&R Fazzolari Pty Ltd v Parramatta City Council* (2009) 254 ALR 1 (Newton and Conolly, 2017).

Conversely, two States provide for a landowner to compel an authority to compulsorily acquire their land, such as where a small parcel would otherwise be retained, being New South Wales (s21–28, *Land Acquisition (Just Terms Compensation) Act 1997*) and Victoria (s76, *Land Acquisition and Compensation Act 1986*) (Brown, 2009).

Further, in the context of public purpose, Hemmings (1997) notes that, generally, in a "before and after" valuation approach to the assessment of compensation, regard may not be had to the public purpose in the 'before' scenario but regard may be had to the public purpose in the 'after' scenario.

3.5.1.1 Broad vs. narrow interpretation of the public purpose

The recent cases of *Goldmate Property Luddenham No 1 Pty Ltd v Transport for NSW* [2024] NSWLEC 39 and the ensuing appeal illustrate the difference between a broad interpretation of the public purpose and a narrow interpretation of the public purpose. The cases concerned the compulsory acquisition of land for the M12 motorway near the Western Sydney Airport at Badgery's Creek.

In *Goldmate Property Luddenham No 1 Pty Ltd v Transport for NSW* [2024] NSWLEC 39, the Court interpreted the public purpose broadly, holding that the public purpose in acquiring the land was for the furtherance of the broader infrastructure plan for the development of the airport and the economic development of western Sydney. Effectively, if the Western Sydney Airport was not being developed, there would be no M12 required. Consequently, as the public purpose was broader than the creation of the M12 motorway, being the broader infrastructure plan, this was to be disregarded and the consequential rezoning of the subject land to a higher order zoning was also to be disregarded in the determination of compensation.

On appeal, the Court took a different view and interpreted the public purpose narrowly (*Goldmate Property Luddenham No 1 Pty Ltd v Transport for NSW* [2024] NSWCA 292), determining that the public purpose was restricted by the scope of the acquiring authority's power to acquire land under the enabling statute (being the *Roads Act* 1993 (NSW)). Under the enabling statute, the acquiring authority was only authorised to acquire land to carry out road work for a freeway and was not empowered to acquire land for broader purposes such as economic development. Consequently, as the public purpose was narrow, being the creation of the M12 motorway only, regard should be given to the broader infrastructure plan and the consequential rezoning of the subject land to a higher order zoning in the determination of compensation.

The Appeal Court enunciated a four step approach, being:

1. the identification of the acquiring authority;
2. the identification, by reference to the empowering legislation, of the public purpose or purposes for which the acquiring authority (identified in (1)) has the power to acquire land;
3. the identification of the acquiring authority's public purpose in acquiring the land, which must fall within the purpose or range of purposes identified in (2) and
4. the determination of the question, which is one of fact, whether there has been any increase or decrease in the value of the land caused by the carrying out of, or the proposal to carry out, the public purpose for which the land was acquired, identified in (3) (any such increase or decrease to be disregarded).

Transport for NSW sought special leave to appeal the decision of the Court of Appeal to the High Court but, in April 2025, special leave was refused. Accordingly, the decision of the Court of Appeal stands.

3.5.2 Disregard of the public purpose

As noted earlier, the public policy intention of the disregard of the public purpose is to ensure landowners receive compensation as if the public purpose did not exist. Therefore, landowners are not penalised if the public purpose decreases the value of their land and they also do not obtain any benefit if the public purpose increases the value of their land.

While disregard of the public purpose is now included in all facilitating statutes, prior to the introduction of the facilitating statutes the principles were enunciated in case law.

3.5.2.1 Pre-facilitating statutes period

Under the *Pointe Gourde* principle, compensation for the compulsory acquisition of land cannot include an increase or decrease in the value of the land which is entirely due to the carrying out, or the proposal to carry out, the scheme underlying the acquisition, being the purpose for which the land was acquired (*Pointe Gourde Quarrying and Transport Co v Sub-Intendent of Crown Lands (Trinidad)* [1947] AC 565) (Brown, 2009).

Further, where, prior to the date of resumption, restrictions may have been placed on the use of the land by Government as steps leading to its resumption and which have a significant effect on the market value of that land resumed (such as zoning for a public purpose), the effect of such restrictions is to be disregarded in the assessment of the market value of that land, with compensation based on that which would otherwise be payable based on the underlying zoning of the land (*Housing Commission of NSW v San Sebastian Pty Ltd* (1978) 37 LGRA 214) (Brown, 2009).

3.5.2.2 Post-facilitating statutes period

Following the introduction of the various facilitating statutes, disregard of the public purpose was codified, with Brown (2009) identifying and listing the references in Commonwealth, State and Territory facilitating statutes (pages 141–142) and Hyam (2014, pages 481–483).

For example, s 56(1) of the NSW *Land Acquisition (Just Terms Compensation) Act 1991* provides that, in the assessment of market value, any increase or decrease in the value of the land caused by the carrying out of, or proposal to carry out, the public purpose for which the land was acquired must be disregarded.

3.5.3 Application by valuers

Applying the principle of the disregard of the public purpose may often be challenging when undertaking a valuation to assess compulsory acquisition compensation. The valuer should carefully and logically consider the subject property and the relevant comparable sales accumulated in the context of the public purpose and the current zoning and look for possible impacts, effects or inter-relationships.

Given the long lead time for many projects for which compulsory acquisition is eventually adopted, the market is aware that the project is coming at some point in time and this may influence the price at which land and properties transact in the local market. Endeavouring to exclude such influences or ascertain what transaction prices may have been in their absence can be both difficult and subjective.

3.6 Summary and conclusions

Focusing on the key issues from a valuer's perspective, Chapters 1 and 2 introduced the concepts and principles of statutory valuation, with statutory valuation for acquisition considered in Chapters 3, 4, 5 and 6. Statutory valuation for rating and taxing is considered in Chapters 7 and 8, with regulatory valuation then considered in Chapters 9 and 10.

Chapter 1 introduced the principles of value in statutory valuation, including the concepts of market value, liberal/conservative estimates of value, highest and best use, potentiality and confirmation of a foresight. This provided the essential foundation principles of statutory valuation that are then applied in the following chapters on statutory valuation.

Chapter 2 considered the practice of statutory valuation including valuation methods with an emphasis on the comparative method of valuation, together with statutory valuation conventions such as the before and after method and the piecemeal method. This provided an understanding of the valuation methods and conventions that may be preferred for application in the following chapters on statutory valuation.

Statutory valuation for acquisition is considered in Chapters 3, 4, 5 and 6. This chapter examined the principles of enabling and facilitating statutes for the compulsory acquisition of land, highlighting the importance for the valuer of ensuring that the public purpose of the acquisition is permitted under the enabling statute.

A generic model of the acquisition process was proposed, identifying the key steps in both acquisition by agreement and compulsory acquisition. The continuous role of negotiation was explored, including after the commencement of acquisition by compulsion, with acquisition by agreement being preferable, where possible, for both the parties and the community who save the costs of litigation. By way of case study, the generic acquisition process was considered in the context of the NSW *Land Acquisition (Just Terms Compensation) Act 1991* and the sequential nature of key dates and time periods considered.

Notions of fairness in the compulsory acquisition process were explored through the concept of *on just terms* and provisions for acquisition due to hardship. Further, the importance of the valuer identifying every interest in the land acquired was considered, as each interest comprises a separate and independent claim for compensation, prior to considering the trade-off between the dispossessed and the community arising from the disregard of the public purpose. This provided an insight into concepts of fairness and equity in statutory valuation in the context of the requirement to disregard the compulsory nature of the acquisition.

Chapter 4 explores the market value bases of compensation in acquisition for which statutory valuation may be required, including market value of the interest and the related market value compensation concepts of severance and injurious affection. This provides a broad understanding of the heads of compensation for acquisition on a value basis that may arise in statutory valuation.

Chapter 5 considers non-market value forms of compensation, including special value, which is noted to rarely arise, payment for solace arising from loss of home which is heavily dependent on the exact wording of the facilitating statute and the concept of disturbance for costs and losses suffered as a result of the acquisition. This provides a broad understanding of the heads of compensation for acquisition that are not necessarily on a value basis but that may arise in statutory valuation.

Chapter 6 concludes the chapters considering statutory valuation for acquisition by examining compensation for other interests such as easements and mortgages, mineral rights and profits à prendre, strata title and substrata, businesses and the emerging statutory valuation areas of native title and indigenous cultural loss. This provides an appreciation of the diversity of interests that may be subject to acquisition and so may require valuation for the assessment of compensation.

Statutory valuation for rating and taxing is considered in Chapters 7 and 8. Chapter 7 introduces the principles of land as the basis for taxation through land tax, stamp duty and rates, considering the general principles of the valuation of land for taxation using the New South

Wales statutory framework as an example. This provides a general understanding of the often arcane and artificial basis of statutory valuation upon which government collects very significant amounts of taxation revenue.

Chapter 8 explores the practice of statutory valuation for rating and taxing through exemptions, concessions and allowances that have evolved over the last century, together with contemporary issues such as mass appraisal, automated valuation models and artificial intelligence, the impact of major disasters on valuation for rating and taxing and the litigation process, including the valuer as an expert witness. This provides an insight into the continued evolution of statutory valuation to address issues as they arise over time, highlighting the importance of a principles-based approach to statutory valuation.

Regulatory valuation, considered in the context of the preparation of valuations for inclusion in financial reporting, is considered in Chapters 9 and 10. Chapter 9 explores the principles of regulatory valuation with regard to the inter-related roles of international accounting standards, international financial reporting standards and international valuation standards, focusing on the primacy of fair value in international financial reporting standards. This provides an understanding of the fundamental conceptual differences and bases between statutory valuation and regulatory valuation.

Finally, Chapter 10 concludes regulatory valuation by exploring the practice of valuation approaches and methods relevant for regulatory valuation such as the income approach and the cost approach and the various valuation methods there within. This provides an appreciation of the wider range of valuation approaches and methods available to the valuer for regulatory valuation, reflecting that, unlike statutory valuation, there is no requirement for defence in Court (negligence claims excepted).

References

API (2021a) *ANZVGP 113 valuations for compensation and compulsory acquisition*, API, Deakin

API (2021b) *ANZVGP 113.1 valuations for compensation and compulsory acquisition resource pack*, API, Deakin

Brown, D (2009) *Land acquisition*, Lexis Nexis Butterworths, Chatswood

Du Plessis, E (2018) How the determination of compensation is influenced by the distinction between the concepts of 'value' and 'compensation', in B Hoops et al (Eds) *Rethinking expropriation law III*, Eleven International Publishing, The Hague

Hemmings, N (1997) Compensation, Chapter in *Valuation principles and practice*, Australian Property Institute, Deakin

Hyam, A (2014) *The law affecting valuation of land in Australia*, The Federation Press, Annandale

Jacobs, M (2015) *Law of compulsory land acquisition*, Thomson Reuters, Pyrmont

Newton, G and Conolly, C (2017) *Land acquisition*, Lexis Nexis Butterworths, Chatswood

Parker, D (2019) Compulsory acquisition compensation issues in Australia, in F Plimmer and W McCluskey (Eds) *Routledge handbook of contemporary issues in expropriation*, Routledge, Abingdon

Parker, D (Ed) (2022) *Principles and practice of property valuation in Australia*, Routledge, Abingdon

Parker, D (2024) Compensation for compulsory acquisition – value vs worth, *Journal of Property Investment and Finance*, Vol 42, No 3, pp315–320

Appendix 1 – Case study

A1.1 Overview

Mickey runs a large horse breeding and racing business, based in Happytown, a small country town in NSW, comprising a 10,400 sqm block of land located on the edge of Happytown together with several hundred hectares of grazing land surrounding Happytown.

Prior to the compulsory acquisition, the 10,400 sqm block of land owned by Mickey was surrounded by Mouse Street, Duck Street and Hound Street, as shown in the before graphic below (2.0). The block comprised a paddock to which horses would be brought from Mickey's surrounding grazing lands for grazing, shoeing, veterinary attention and general care, together with a homestead and a water bore, a meadow used for grazing horses with a billboard, a blacksmith's shop and two houses, each of which is separately fenced off, with the entire 10,400 sqm being zoned for medium density residential development.

The proposed acquisition will facilitate the construction, operation and maintenance of the Happytown Freeway through Mickey's block of land, as shown in the after graphic below (3.0). The land to be acquired is marked by an arrow in the graphic below. The acquisition will be gazetted on 31 December 2024.

As a result of the compulsory acquisition, the portions of the block retained by Mickey north and south of the proposed Happytown Freeway will no longer meet Council's land size criteria for medium density residential development but will meet Council's criteria for low density residential development.

A1.2 Before graphic

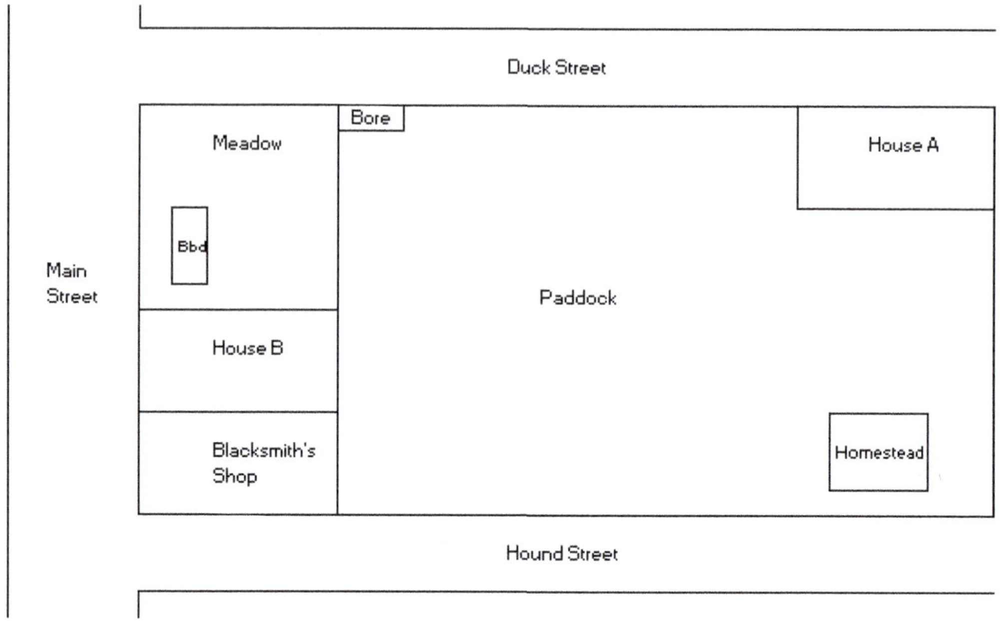

A1.3 After graphic

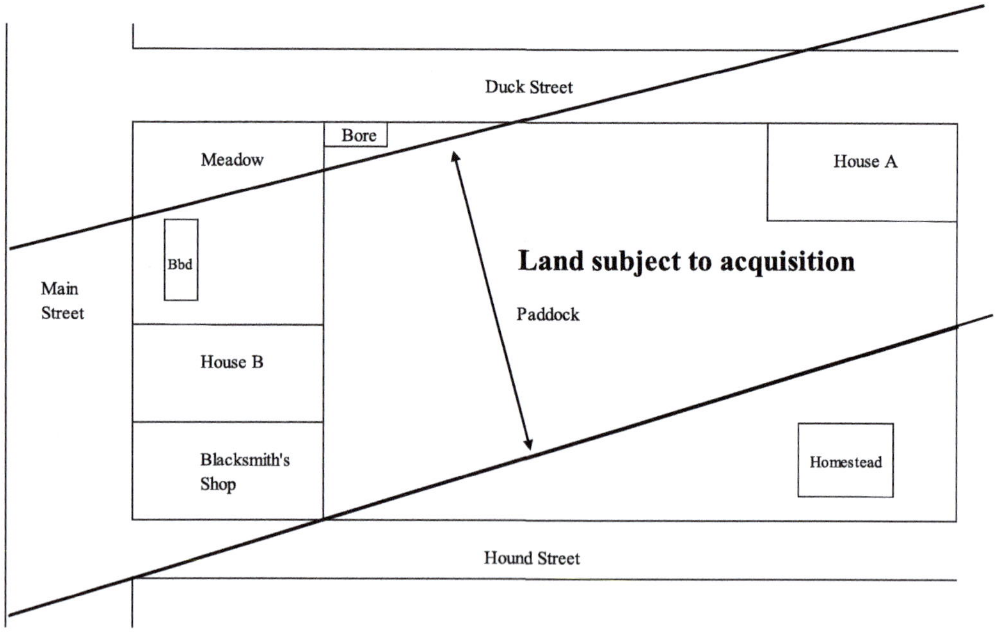

A1.4 Land areas

The respective land areas are as follows:

Area of Mickey's block	10,400 sqm
Area of homestead and paddock	7,400 sqm
Area of shop site	600 sqm
Area of house A and B sites	600 sqm each
Area of meadow	1,200 sqm
Area to be acquired	7,800 sqm
Comprising:	
Paddock	5,250 sqm
House A	600 sqm
House B	600 sqm
Shop	600 sqm
Meadow	750 sqm
Portion of the block retained north of the proposed freeway	600 sqm (450 sqm meadow, 150 sqm paddock)
Portion of the block retained south of the proposed freeway	2,000 sqm

A1.5 Parties

Mickey occupies the homestead and the paddock.

Mickey leases house A to Minnie on a 5-year lease expiring 31 December 2025 at $100 pcm. Minnie has been the occupant of house A as her principal place of residence for over a decade. Minnie is very elderly and moving house is a very traumatic event causing extreme anxiety. The acquiring authority requires possession on 31 December 2024.

Mickey leases house B to Daisy on a month-to-month tenancy at $100 pcm. Daisy has been the occupant of house B as her principal place of residence for 3 years. Daisy was thinking of moving to a smaller residence and welcomes the opportunity to relocate. The acquiring authority requires possession on 31 December 2024.

Mickey leases the shop to Pluto on a ten-year lease expiring 31 December 2025 at $250 pcm. Pluto runs a blacksmith's shop, with existing use rights, which services Mickey's horses in the paddock as well as minor metalwork jobs for residents of Happytown, showing future maintainable profits of $1,000 pa and holding stock worth $5,000. Pluto has decided it is time to retire and will not relocate his blacksmith's shop elsewhere following the compulsory acquisition.

Mickey has given a licence to Goofy to use the meadow for 6 months expiring 31 December 2024 at $1 pcm.

Mickey owns the billboard and has given a tenancy at will to Donald to use the billboard at $60 pcm.

A1.6 Facilitating statute

The proposed acquisition notice stated that the land was to be acquired *for the purposes of the Roads Act . . . in connection with the construction, operation and maintenance of the Happy-town Freeway.*

A1.7 Valuation information

Your accumulation, analysis, adjustment and application of comparable sale transactions for houses A and B indicate that the market value for each house and curtilage is $250,000.

Your accumulation, analysis, adjustment and application of comparable rental transactions for houses A and B indicate that the market rental for each house is $500 pcm.

Your accumulation, analysis, adjustment and application of comparable sale transactions for short-term residential leasehold interests indicate that such interests trade on a capitalisation rate of 10%.

Your accumulation, analysis, adjustment and application of comparable rental transactions for the blacksmith's shop indicate that the market rental for the shop is $750 pcm.

Your accumulation, analysis, adjustment and application of comparable sales for freehold shops and curtilage indicate that shops trade on a capitalisation rate of 5%.

Your accumulation, analysis, adjustment and application of comparable sale transactions for short-term leasehold shop interests indicate that such interests trade on a capitalisation rate of 10%.

Your analysis of the turnover of the blacksmith's shop shows future maintainable profits of $1,000 pa.

Your accumulation, analysis, adjustment and application of comparable business sales show that blacksmith's businesses trade on a capitalisation rate of 10%.

Your accumulation, analysis, adjustment and application of comparable rentals for meadow grazing indicate a market rental for 1,200 sqm to be $120 pcm.

Your accumulation, analysis, adjustment and application of comparable rentals for billboards indicate a market rental to be $120 pcm.

Your accumulation, analysis, adjustment and application of comparable sales for billboard businesses indicate that billboard businesses trade on a capitalisation rate of 10%.

Your analysis of the turnover of the horse breeding and training business conducted on the land shows future maintainable profits of $10,000 pa.

Your accumulation, analysis, adjustment and application of comparable business sales show that horse breeding and training businesses trade on a capitalisation rate of 15%.

Your accumulation, analysis, adjustment and application of comparable sales for pastoral land with a water supply suitable for grazing indicate a market rate of $100 psm.

Your accumulation, analysis, adjustment and application of comparable sales for low density residential development land indicate a market rate of $250 psm.

Your accumulation, analysis, adjustment and application of comparable sales for medium density residential development land indicate a market rate of $500 psm.

You have received advice that the replacement cost of the billboard is $250,000 and that the existing billboard has a potential remaining usable life of 10 years, having been erected 40 years ago.

A1.8 Disturbance claims

Mickey has submitted four invoices for costs associated with the compulsory acquisition, being $10,000 for legal advice on the acquisition, $5,000 for legal advice on likely success in an action in the Land and Environment Court, $5,000 for valuer's fees and $13,838 for stamp duty on two new houses.

Minnie has submitted three invoices for costs associated with vacating house A, being $100 for a solicitor to review the lease of her new rental house, $500 for removalists and $1,000 for her friend who helped Minnie find a new rental house to move into.

Daisy has submitted three invoices for costs associated with vacating house B, being $100 for a solicitor to review the lease of her new rental house, $500 for removalists and $2,500 for the cost of repainting the interior of her new rental house.

Pluto has submitted four invoices for costs associated with vacating the blacksmith's shop, being $500 for the hire of a van to move stock from the shop to his home garage, $150 for an advice of retirement sign, $1,000 for losses suffered in a closing down sale and $500 for setting up his business landline, fax machine and internet connection in his home.

Donald has submitted one invoice for $10,000 for loss of profits from billboard advertising for a period of 12 months after the date of acquisition.

Goofy has submitted one invoice for the costs associated with moving his horse from the meadow to another meadow, which is within walking distance, being $50,000.

4 Compulsory acquisition

Compensation for market value, severance and injurious affection

4.1 Introduction

This book is an introduction to two very different valuation worlds in Australia – the complex and often arcane national realm of statutory valuation and the dynamic global empire of regulatory valuation. Statutory valuation is different from regulatory valuation, with the former being based on relevant applicable statutes and being capable of challenge on appeal in Court while the latter is based on relevant applicable regulations and incapable of challenge in Court (negligence claims excepted).

As an introductory textbook on statutory and regulatory valuation, this book is a companion to the Australian valuation textbooks, *Introduction to Property Valuation in Australia* (Parker, 2024) and *Principles and Practice of Property Valuation in Australia* (Parker, 2022), which provide a wider analysis of key principles underlying property valuation and current techniques and issues in the practice of property valuation for major sectors of the Australian property market.

Statutory valuation in Australia is most commonly encountered in valuations for rating, taxing and acquisition but may arise wherever a statute prescribes a basis of or approach to valuation. Regulatory valuation is most commonly encountered in valuations for financial statements but may arise wherever a regulation prescribes a basis of or approach to valuation.

Rating and taxing of land raise significant amounts of revenue for State Governments and Councils in Australia, based on statutory valuation. For example, the NSW 2025–26 Budget Statement estimates land tax revenue for 2024–2025 to be $8.2 billion with the City of Brisbane Annual Plan and Budget 2024–2025 estimating proposed rates and utility charges to be $1.51 billion.

Acquisition of land, property and businesses commonly arises for major infrastructure projects and may be by agreement or by compulsion, based on statutory valuation. For example, such major infrastructure projects in Australia currently include the multi-billion-dollar Sydney Metro, Melbourne Suburban Rail Loop and Melbourne to Brisbane Inland Rail projects.

Valuation of land and property holdings based on regulatory valuation arises for all reporting entities in Australia that publish financial statements upon which it is reasonable to expect that users will rely for decision making. For example, at June 2025 there are over 2,000 reporting entities listed on the Australian Stock Exchange which each publish financial statements, with the financial statements for Goodman Group, as at 30 June 2024, showing the fair value of total property assets to be $20.2 billion, based on regulatory valuation.

Focusing on the key issues from a valuer's perspective, the concepts and principles of statutory valuation are introduced in Chapters 1 and 2, with statutory valuation for acquisition considered in Chapters 3, 4, 5 and 6 and statutory valuation for rating and taxing considered in Chapters 7 and 8. Regulatory valuation is then considered in Chapters 9 and 10.

DOI: 10.1201/9781032616957-4

Chapter 1 introduced the principles of value in statutory valuation, including the concepts of market value, liberal/conservative estimates of value, highest and best use, potentiality and confirmation of a foresight. This provided the essential foundation principles of statutory valuation that are then applied in the following chapters on statutory valuation.

Chapter 2 considered the practice of statutory valuation including valuation methods for statutory valuation with an emphasis on the comparative method of valuation, together with statutory valuation conventions such as the before and after method and the piecemeal method. This provided an understanding of the valuation methods and conventions that may be preferred in the following chapters on statutory valuation.

Statutory valuation for land acquisition is considered in chapters 3, 4, 5 and 6, supported by an evolving case study in Appendix 1 of each chapter, including the principal bases of compensation being:

- compensation for market value for the interest acquired;
- compensation for severance;
- compensation for injurious affection;
- compensation for special value;
- compensation for solatium or disadvantage resulting from relocation and
- compensation for disturbance.

Chapter 3 introduced the principles of statutory valuation for land acquisition, including the statutory framework and process, with a focus on the concept of disregard of the scheme in compensation assessment and the distinction between concepts of value and worth. This provided an insight into concepts of fairness and equity in statutory valuation in the context of the requirement to disregard the compulsory nature of the acquisition.

This chapter explores the market value bases of compensation in acquisition for which statutory valuation may be required, including market value of the interest and the related market value compensation concepts of severance and injurious affection. This provides a broad understanding of the heads of compensation for acquisition on a value basis that may arise in statutory valuation.

Chapter 5 considers non-market value forms of compensation including special value, which is noted to rarely arise, payment for solace arising from loss of the primary place of residence and the concept of disturbance for costs and losses suffered as a result of the acquisition. This provides a broad understanding of the heads of compensation for acquisition that are not necessarily on a value basis but which may arise in statutory valuation.

Chapter 6 concludes the chapters considering statutory valuation for acquisition by examining compensation for other interests such as easements and mortgages, mineral rights and profits à prendre, strata title and substrata, businesses and the emerging statutory valuation areas of native title and indigenous cultural loss. This provides an appreciation of the diversity of interests that may be subject to acquisition and so may require valuation for the assessment of compensation.

Statutory valuation for rating and taxing is considered in Chapters 7 and 8. Chapter 7 introduces the principles of land as the basis for taxation through land tax, stamp duty and rates, considering the general principles of the valuation of land for taxation using the New South Wales statutory framework as an example. This provides a general understanding of the often arcane and artificial basis of statutory valuation upon which Government collects very significant amounts of taxation revenue.

Chapter 8 explores the practice of statutory valuation for rating and taxing through exemptions, concessions and allowances that have evolved over the last century, together with contemporary issues such as mass appraisal, automated valuation models and artificial intelligence, the impact of major disasters on valuation for rating and taxing and the litigation process including the valuer as an expert witness. This provides an insight into the continued evolution of statutory valuation to address issues as they arise over time, highlighting the importance of a principles-based approach to statutory valuation.

Regulatory valuation, considered in the context of the preparation of valuations for inclusion in financial reporting, is considered in Chapters 9 and 10. Chapter 9 explores the principles of regulatory valuation with regard to the inter-related roles of international accounting standards, international financial reporting standards and international valuation standards, focusing on the primacy of fair value in international financial reporting standards. This provides an understanding of the fundamental conceptual differences and bases between statutory valuation and regulatory valuation.

Finally, Chapter 10 concludes regulatory valuation by exploring the practice of valuation approaches and methods relevant for regulatory valuation such as the income approach and the cost approach and the various valuation methods there within. This provides an appreciation of the wider range of valuation approaches and methods available to the valuer for regulatory valuation reflecting that, unlike statutory valuation, there is no requirement for defence in Court (negligence claims excepted).

Accordingly, this chapter will now consider:

- compensation for market value;
- compensation for severance and
- compensation for injurious affection or betterment.

4.2 Compensation for market value – principles

This section seeks to guide the valuer through the valuation aspects of the key issues concerning compensation for market value from a valuer's viewpoint. Greater detail on the legal aspects of the key issues may be found in Jacobs (2015), Hyam (2014), Brown (2009) and Newton and Conolly (2017).

Section 3.2.2.2 in Chapter 3 noted that the acquiring authority will provide the landowner with a formal notice advising of the proposed acquisition, the reason for the acquisition (generally referred to as the *public purpose*) and the relevant enabling statute.

The acquiring authority may be required to publish the formal notice in a government publication (such as the Government Gazette). At the point of publication, ownership of the land and property transfers to the acquiring authority and the landowner's interest converts to a claim for compensation, with the date of publication usually being the date of valuation.

Significantly, compensation is in the form of money and not in the form of another property. While, in residential claims, the dispossessed often say that they would prefer an alternative house of the same size and quality close to their relevant amenities, the notion that a party is entitled to be reinstated in an equivalent property is not part of the statutory framework. Consistently, the heterogeneity of property means that such an alternative house cannot practically be provided (the kitchen will be superior/inferior to the subject, the block will be larger/smaller than the subject, etc.).

However, in theory, for the purposes of residential property, the use of similar properties that sold in the local area as comparable sales should result in a valuation at an amount that would allow the dispossessed to buy such a property and relocate within the local area.

The facilitating statutes specify the forms of compensation payable, with all providing for compensation for all interests acquired. Some of those interests acquired may include interests in land and buildings, with the compensation for interests in land and buildings generally being some form of market value as defined in the facilitating statute.

4.2.1 Facilitating statutes

In many cases, compensation for the market value of the interest acquired forms the largest portion of the compensation received by the dispossessed. The exact definition and/or approach to market value in each facilitating statute is, therefore, very important for the valuer as it determines the basis of the (usually) largest portion of compensation.

Market value is variously defined in the facilitating statutes as follows:

Commonwealth: *Lands Acquisition Act* 1989

Section 55(2)(a)(i):

(2) In assessing the amount of compensation to which the person is entitled, regard shall be had to all relevant matters, including:

(a) except in a case to which paragraph (b) applies:

(i) the market value of the interest on the day of the acquisition.

Section 56:

For the purposes of this Division, the market value of an interest in land at a particular time is the amount that would have been paid for the interest if it had been sold at that time by a willing but not anxious seller to a willing but not anxious buyer.

New South Wales: *Land Acquisition (Just Terms Compensation) Act* 1991

Section 56:

(1) *In this Act:*

"market value" of land at any time means the amount that would have been paid for the land if it had been sold at that time by a willing but not anxious seller to a willing but not anxious buyer, disregarding (for the purpose of determining the amount that would have been paid):

(a) any increase or decrease in the value of the land caused by the carrying out of, or the proposal to carry out, the public purpose for which the land was acquired, and
(b) any increase in the value of the land caused by the carrying out by the authority of the State, before the land is acquired, of improvements for the public purpose for which the land is to be acquired, and
(c) any increase in the value of the land caused by its use in a manner or for a purpose contrary to law.

(2) When assessing the market value of land for the purpose of paying compensation to a number of former owners of the land, the sum of the market values of each interest in the land must not (except with the approval of the Minister responsible for the authority of the State) exceed the market value of the land at the date of acquisition.

Victoria: *Land Acquisition and Compensation Act* 1986

Section 40:

"market value", in relation to any interest in land on a particular date, means the amount of money that would have been paid for that interest if it had been sold on that date by a willing but not anxious seller to a willing but not anxious purchaser.

Australian Capital Territory: *Lands Acquisition Act* 1994

Section 45(1) and (2)(a)(i):

(1) The amount of compensation to which a person is entitled under this part in respect of the acquisition of an interest in land is such amount as, having regard to all relevant matters, will justly compensate the person for the acquisition.
(2) In assessing the amount of compensation to which the person is entitled, regard shall be had to all relevant matters, including –

(a) except in a case to which paragraph (b) applies –

(i) the market value of the interest on the day of acquisition . . .

Section 46:

For this division, the "market value" of an interest in land at a particular time is the amount that would have been paid for the interest if it had been sold at that time by a willing but not anxious seller to a willing but not anxious buyer.

Northern Territory: *Land Acquisition Act* 1978

Section 66(1) as read with Sch 2, r 2(a):

(1) Subject to this section, in assessing compensation under this Act, the Tribunal must have regard to, but is not bound by, the rules set out in Schedule 2.

Sch 2, r 2(a):

. . . the consideration that would have been paid for the land if it had been sold on the open market on the date of acquisition by a willing but not anxious seller to a willing but not anxious buyer.

Queensland: *Acquisition of Land Act* 1967

Section 20(1) and (2):

(1) In assessing the compensation to be paid, regard shall in every case be had not only to the value of land taken but also:. . .
(2) Compensation shall be assessed according to the value of the estate or interest of the claimant in the land taken on the date when it was taken.

South Australia: *Land Acquisition Act* 1969

Section 25(1)(a) and (b):

(1) The compensation payable under this Act in respect of the acquisition of land shall be determined according to the following principles:

(a) the compensation payable to a claimant shall be such as adequately to compensate him for any loss that he has suffered by reason of the acquisition of the land; and
(b) in assessing the amount referred to in paragraph (a) of this section consideration may be given to –

(i) the actual value of the subject land; and
(ii) the loss occasioned by reason of severance, disturbance or injurious affection.

Tasmania: *Land Acquisition Act* 1993

Section 27(1)(a):

(1) In determining compensation under this Act, regard is to be had to the following matters:

(a) the market value of the estate of the claimant in the subject land.

Western Australia: *Land Administration Act* 1997

Section 202(1) and (3):

(1) Every person having any interest in land which is taken under *Part 9* is entitled, subject to this Part, to compensation for the interest from the acquiring authority.

(3) No compensation is payable under this section for land transferred under *section 75* other than in respect of –

(a) lawful improvements made to the land since the transfer; or
(b) consideration paid for the transfer of the land.

The Commonwealth, New South Wales, Victoria, Australian Capital Territory and the Northern Territory each adopt market value and the hypothetical willing buyer/seller principles from *Spencer*, with Queensland, South Australia and Tasmania adopting value/market value only and Western Australia adopting consideration.

As considered in Chapters 1 and 2:

- market value usually follows a normative definition (being what a hypothetical purchaser and hypothetical vendor would agree) and does not provide any concept of compensation for worth to the dispossessed (commonly called value to the owner);
- market value disregards the public purpose;
- market value disregards uses contrary to law (such as land and buildings associated with the growth, development or distribution of illegal drugs);
- market value for dispossession is to be a liberal estimate;

- market value is based on the concept of highest and best use;
- market value has regard to potentiality;
- market value may confirm a foresight;
- market value is determined as at the date of acquisition; with
- market value preferably determined by the comparative method of valuation.

However, generally, if compensation for market value is on the basis of a higher and better use than the current use, this precludes a claim for disturbance to the extent that it is inconsistent with a claim for market value based on the potential use of the land, reflecting that, in realising highest and best use value, current use would be discarded.

4.2.2 Relocation and reinstatement

On acquisition, the holder of an interest to which a business is attached may choose to:

- relocate and continue trading – in which case compensation for business relocation may be payable as considered below and in Section 5.4.1 of Chapter 5 or
- cease trading and close down – in which case compensation for business extinguishment may be payable, being the value of the business as a going concern, which is considered in Section 4.2.3 and Section 6.4 of Chapter 6.

Compensation for relocation and reinstatement of a business is a complex area of statutory valuation, arising where a business is attached to the interest in land that is compulsorily acquired. Where a business is being carried out by the dispossessed, the value of the land will be determined on the highest and best use and that may include the value generated by the net maintainable earnings of the business that could be carried out on the land.

In *Hua v Hurstville City Council* [2010] NSWLEC 61, the resumption of a bakery business led to the consideration of reinstatement and relocation, with the Court noting that relocation involved moving something to a different location (effectively, relocating) whereas reinstatement involved returning something to its former position or state (effectively, recreating).

Costs associated with business relocation may be addressed by compensation for disturbance, which is considered in Section 5.4.1 in Chapter 5.

Costs associated with reinstatement may arise where the acquired land is used for a purpose for which there is no general market (such as an equestrian centre or a suburban heliport), the determination of market value through the application of the comparative method of valuation is challenging and the dispossessed may seek compensation to recreate the property on an identified block of land.

Jacobs (2015) quotes Brown (1991):

> It is commonly said that there are three requirements to be met for the principle of reinstatement to apply as a means of valuation:
>
> (a) the claimant's acquired land must be devoted to a purpose of such a nature that there is no general demand for that purpose;
> (b) the claimant's acquired land must have continued to be devoted for that purpose, but for the acquisition; and
> (c) the claimant must have a bona fide intention to reinstate himself in some other place for the same purpose.
>
> (pages 686–687)

Effectively, the dispossessed has to evidence a genuine intention to reinstate the business rather than simply asserting an intention to do so. However, the reinstated premises must be reasonably comparable to the dispossessed premises and not be superior to that which was acquired (*Commissioner of Highways v Shipp Bros Pty Ltd* (1978) 43 LGRA 355), with allowance made if there is no alternative other than to relocate to superior premises (*Keogh v Housing Commission of Victoria (No 2)* (1969) 18 LGRA 295) and with the dispossessed expected to mitigate their loss where possible.

In only very limited circumstances is market value assessed by reference to reinstatement. The facilitating statutes that permit these limited situations, with varying conditions, include:

- Commonwealth – *Land Acquisition Act* 1989, s58(1)(e);
- New South Wales – *Land Acquisition (Just Terms Compensation)*

  Act 1991, s56(3);

- Victoria – *Land Acquisition and Compensation Act* 1986, s42;
- South Australia – *Land Acquisition Act* 1969, s 25(1)(i);
- Tasmania – *Land Acquisition Act* 1993, s31;
- Northern Territory – *Lands Acquisition Act* 1978, s63 and s66 read with Sch 2 rr 7–9;
- Australian Capital Territory – *Lands Acquisition Act* 1994, ss 48 and 51 (Brown, 2009)

with no specific provision existing in Queensland (Jacobs, 2015).

4.2.3 Extinguishment of business

On acquisition, a business which is attached to the interest in land may choose to cease trading and close down, in which case compensation for business extinguishment may be payable, being the value of the business as a going concern which is considered below and in Section 6.4 in Chapter 6.

Compensation for the extinguishment of a business is a complex area of statutory valuation, arising where a business is attached to the interest in land that is compulsorily acquired. Where a business is being carried out by the dispossessed, the value of the land will be determined on the highest and best use and that may include the value generated by the net maintainable earnings of the business that could be carried out on the land.

Brown (2009) notes that the facilitating statutes do not directly refer to situations where the act of resumption causes, or partly contributes to, the extinguishment of a business, with the dispossessed often needing to rely on the provisions governing disturbance, loss or damage.

If, for example, where the compulsory acquisition of a baker's shop includes the acquisition of the leasehold interest in the shop held by the baker and the extinguishment of the business run by the baker attached to the leasehold interest, compensation may be by market value for both the leasehold interest and the business interest or may be by market value for the leasehold interest and disturbance compensation for the value of the bakery business as a going concern, being further addressed in Section 6.4 of Chapter 6.

Compensation for business extinguishment and business relocation may require consideration on a relative basis. In *Re Mills and Minister for Finance and Administration* (2002) 70 ALD 780, the Court held that if the costs of relocation exceeded the costs of extinguishment, compensation should be based on the cost of extinguishment. However, in *Director of Buildings and Lands v Shun Fung Ironworks Ltd* [1995] 1 All ER 846, the Court noted that *this will not always be so, and a rigid limitation . . . could lead to injustice* (853).

4.3 Compensation for market value – practice

Appendix 1 contains a case study which will be used to illustrate the various aspects of compulsory acquisition considered in this chapter.

4.3.1 Facilitating statute

As the subject property is in New South Wales, the relevant facilitating statute is the *Land Acquisition (Just Terms Compensation) Act 1991*, which prescribes compensation for market value, stating at s56:

> (4) In this Act:
> "market value" of land at any time means the amount that would have been paid for the land if it had been sold at that time by a willing but not anxious seller to a willing but not anxious buyer, disregarding (for the purpose of determining the amount that would have been paid):
>
> (d) any increase or decrease in the value of the land caused by the carrying out of, or the proposal to carry out, the public purpose for which the land was acquired, and
> (e) any increase in the value of the land caused by the carrying out by the authority of the State, before the land is acquired, of improvements for the public purpose for which the land is to be acquired, and
> (f) any increase in the value of the land caused by its use in a manner or for a purpose contrary to law.

Accordingly, the identified interests in land may be eligible for market value compensation disregarding the public purpose, the benefit of any prior works by the acquiring authority and any uses contrary to law, each of which should be considered separately.

Following the case study in Appendix 1, as noted in Section 3.2.5 of Chapter 3, the proposed acquisition notice stated that the land was to be acquired *for the purposes of the Roads Act . . . in connection with the construction, operation and maintenance of the Happytown Freeway.* Accordingly, the public purpose does not appear from the case study to have any impact on the subject property that would require disregard. Further, the case study does not indicate any prior works by the acquiring authority that should be disregarded.

The lawfulness of each of the uses should be checked against the relevant planning instruments to confirm that the homestead, houses A and B, the shop and the billboard each have development approval or existing use rights and that the uses for residential, commercial, paddock and grazing are each permitted under the relevant zoning. The case study does not indicate any unlawful uses.

4.3.2 Interests in land

In Section 3.4.2 of Chapter 3, a matrix was constructed to identify the range of parties and the range of interests in land from the case study, as restated in Table 4.1.

Accordingly, the market value of each of the following acquired interests should be assessed to determine the compensation payable:

- Mickey – freehold of all properties and paddock and business;
- Minnie – leasehold of house (lease);

Table 4.1 Case study – parties and interests in land

Party	*Freehold*	*Leasehold*	*Lic/TcyWill*	*Rem/Rev*	*Mortgage*	*Easement*	*Business*
Mickey	√						√
Minnie		√					
Donald			√				√
Daisy		√					
Pluto		√					√
Goofy			√				

Source: Author

- Daisy – leasehold of house (periodic tenancy);
- Donald – tenancy at will of billboard and business;
- Pluto – leasehold of shop (lease) and business and
- Goofy – licence of meadow.

The case study notes that *The acquisition will be gazetted on 31 December 2024*, such that the date of valuation is 31 December 2024.

4.3.3 Compensation for market value

The market value compensation for each interest in land may be considered as follows:

4.3.3.1 Mickey

Mickey holds the following freehold interests in land that may be compensated by market value:

- house A;
- house B;
- shop;
- meadow;
- billboard licence and
- paddock.

together with the billboard structure.

Your accumulation, analysis, adjustment and application of comparable sale transactions for houses A and B indicate that the market value for each house and curtilage is $250,000.

For house A, the lease to Minnie expires 12 months after the acquisition date such that the market value should reflect the delay in obtaining vacant possession. Ideally, comparable sales of similar houses with deferred vacant possession would be accumulated, analysed, adjusted and applied. However, such sales are likely to be rare so a subjective judgmentally based adjustment may be required to the comparative method of valuation, such as the deduction of a capital sum, percentage or the rounding down of the vacant possession value.

Your accumulation, analysis, adjustment and application of comparable rental transactions for houses A and B indicate that the market rental for each house is $500 pcm.

Your accumulation, analysis, adjustment and application of comparable sale transactions for short-term residential leasehold interests indicate that such interests trade on a capitalisation rate of 10%.

For house B, the month-to-month tenancy to Daisy is a periodic tenancy and so can be terminated on one month's notice and the house may be sold with vacant possession and so valued using the comparative method of valuation.

The lease of the shop to Pluto expires 12 months after the acquisition date and so the shop may be valued using the capitalisation of income method if there were no comparable sales of similarly leased shops.

Your accumulation, analysis, adjustment and application of comparable rental transactions for the blacksmith's shop indicate that the market rental for the shop is $750 pcm.

Your accumulation, analysis, adjustment and application of comparable sales for freehold shops and curtilage indicate that shops trade on a capitalisation rate of 5%.

Your accumulation, analysis, adjustment and application of comparable sale transactions for short-term leasehold shop interests indicate that such interests trade on a capitalisation rate of 10%.

The licence to Goofy for the use of the meadow expires on the acquisition date and so the meadow may be sold with vacant possession and valued using the comparative method of valuation.

The meadow is zoned for medium density residential development and your accumulation, analysis, adjustment and application of comparable sales for medium density residential development land indicate a market rate of $500 psm and the land may be valued using the comparative method of valuation.

The tenancy at will to Donald for the billboard may be terminated at any date and so could be relet at the market rental and sold as an investment. In the absence of comparable sales of billboards as investments, the capitalisation of income method of valuation could be used to determine market value.

Your accumulation, analysis, adjustment and application of comparable rentals for billboards indicate a market rental to be $120 pcm.

Your accumulation, analysis, adjustment and application of comparable sales for billboard businesses indicate that billboard businesses trade on a capitalisation rate of 10%.

You have received advice that the replacement cost of the billboard is $250,000 and that the existing billboard has a potential remaining usable life of 10 years, having been erected 40 years ago. The depreciated replacement cost method of valuation would be appropriate to determine the value of the billboard structure.

As there is the potential for double counting if the billboard is compensated both as an income generating investment and as a physical billboard, compensation will only be payable on one basis. Given the relative value as an investment and as a physical asset, Mickey may be likely to choose to be compensated for the loss of the physical asset.

The paddock land is zoned for medium density residential development and your accumulation, analysis, adjustment and application of comparable sales for medium density residential development land indicate a market rate of $500 psm and the land may be valued using the comparative method of valuation.

The market value of Mickey's interests in land is summarised in Table 4.2.

Table 4.2 Mickey – market value of freehold interest

Interest in land	*Method of valuation*		*Market value*
House A and curtilage	Comparative method	Rounded down to reflect 6 month unexpired lease – say less $10,000	$240,000
House B and curtilage	Comparative method		$250,000
Shop and curtilage	Capitalisation of income method	$3,000 pa × 5% for 1 year + $9,000 pa in perp deferred 1 year at 5%	$175,000
Meadow	Comparative method	750 sqm × $500 psm	$375,000
Billboard structure	Depreciated replacement cost	$250,000 × 80% deprecation	$50,000
Paddock	Comparative method	5,250 sqm × $500 psm	$2,625,000
Total			**$3,715,000**

Source: Author

4.3.3.2 Minnie

Minnie holds a leasehold interest in house A that may be compensated by market value. As a leasehold interest, it is the present value of the profit rent for 12 months that requires determination:

Market rent	$500 pcm
Less current rent	$100 pcm
Profit rent	$400 pcm or $4,800 pa
Present value 10%	0.9090909
Market value	$4,400

4.3.3.3 Daisy

Daisy holds a leasehold interest (periodic tenancy) in house B that may be compensated by market value.

As the periodic tenancy to Daisy is not transferable and can be terminated on one month's notice, no profit rent arises and so Daisy's interest has no market value.

4.3.3.4 Pluto

Pluto holds a leasehold interest (lease) in the blacksmith's shop that may be compensated by market value. As a leasehold interest, it is the present value of the profit rent for 12 months that requires determination:

Market rent	$750 pcm
Less current rent	$250 pcm
Profit rent	$500 pcm or $6,000 pa

Present value 10%	0.9090909
Market value	$5,500

The blacksmith's business may have an entitlement to market value compensation or to disturbance compensation under the *Land Acquisition (Just Terms Compensation) Act* 1991 (see Section 6.4.3.2.2, Chapter 6).

4.3.3.5 Goofy

Goofy holds a leasehold interest (licence) in the meadow that may be compensated by market value.

As the licence to Goofy expires on the acquisition date, no profit rent arises and so Goofy's interest has no market value.

4.3.3.6 Donald

Donald holds a leasehold interest (tenancy at will) in the billboard that may be compensated by market value.

As the tenancy at will to Donald is not transferable and can be terminated on one month's notice, no profit rent arises and so Donald's interest has no market value.

4.3.3.7 Compensation for market value – summary

The market value compensation is summarised in Table 4.3.

Table 4.3 Summary – market value compensation

Dispossessed	*Interest in land*		*Market value*
Mickey	Freehold – house A	$240,000	
	Freehold – house B	$250,000	
	Freehold – shop	$175,000	
	Freehold – meadow	$375,000	
	Billboard structure	$50,000	
	Freehold – paddock	$2,625,000	$3,715,000
Minnie	Leasehold – house A		$4,400
Daisy	Leasehold (periodic tenancy) – house B		$0
Pluto	Leasehold – shop		$5,500
Goofy	Leasehold (licence) – meadow		$0
Donald	Leasehold (tenancy at will) – billboard		$0
Total			**$3,724,900**

Source: Author

4.4 Compensation for severance

Compensation for severance arises where an interest in land is severed into two or more pieces following the acquisition of land by an acquiring authority. Without an acquisition of land severing a parent parcel, there can be no grounds for compensation for severance. A common example is a large rural landholding which is divided into two separate sections following the acquisition of a strip of land running through the middle of the block.

4.4.1 Compensation for severance – principles

Newton and Conolly (2017) note that severance is variously defined in the facilitating statutes as follows:

Commonwealth: *Lands Acquisition Act* 1989
Section 55(2)(a)(iii) provides for regard to be had to *any reduction in market value* caused *by the severance by the acquisition of the acquired interest from the other interest.*

New South Wales: *Land Acquisition (Just Terms Compensation) Act* 1991
Section 55(c) includes any loss attributable to severance as compensable with s55(f) addressing any increase or decrease in the value of the owner's other land which adjoins or is severed from the acquired land. Section 58 defines loss attributable to severance as *the amount of any reduction in the market value of any other land of the person entitled to compensation which is caused by that other land being severed from other land of that person.*

Victoria: *Land Acquisition and Compensation Act* 1986
Section 41(1)(c) includes *any loss attributable to severance* as compensable, with s40 defining loss attributable to severance as *the amount of any reduction in the market value of any other interest of the claimant in other land used in conjunction with the acquired land which is caused by its severance from the acquired land.*

Australian Capital Territory: *Lands Acquisition Act* 1994
Section 45 is similar to the Commonwealth provision.

Northern Territory: *Land Acquisition Act* 1978
Sch 2 r 2(c) provides that, in assessing compensation, the Tribunal may take into account *the amount of any reduction in the value of other land of the claimant caused by its severance from the acquired land by the acquisition.*

Queensland: *Acquisition of Land Act* 1967
Section 20(1)(a) provides that in assessing compensation regard shall be had to the *damage (if any) caused by . . . the severing of the land taken from other land of the claimant* with s20(3) addressing set off for enhancement.

South Australia: *Land Acquisition Act* 1969
Section 25(1)(b)(ii) provides that in assessing compensation consideration may be given to the *loss occasioned by reason of severance* with s25(1)(j) addressing set off for enhancement.

Tasmania: *Land Acquisition Act* 1993
Section 27(1)(c) provides that in determining compensation regard shall be had to *the damage caused by severance of the subject land from other land belonging to the claimant.*

Western Australia: *Land Administration Act* 1997
Section 241(7) refers to *adjoining land* and damage suffered due to the severing of the land taken or due to a reduction of the value of that adjoining land.

While there is commonality in intent in the various facilitating statutes concerning severance, there are differences in language such that the valuer should carefully consider the relevant statute given the location of the acquired land and the impact of the severance on the retained land.

Concerning severance, Jacobs (2015) notes that the relevant inquiry is whether or not there is diminution in the value of the remaining land caused by the resumption, depending on the specific facts of the case.

An example of severance arose in *Benton v Road Construction Authority (No 2)* [1992] 2 VR 495, where the severance resulted in the need for the owner to travel extra distances in respect of established procedures for hay carting and livestock droving, which was held to be a compensable loss (Newton and Conolly, 2017).

Hyam (2014) notes that a dispossessed owner may claim severance of the acquired land from remaining land of the owner even though the lands were not contiguous, such as parcels separated by streets (*Crisp & Gunn Co-operative Ltd v Hobart Corporation* (1963) 9 LGRA 378) though Jacobs (2015) notes the need for there to be a *connecting factor* (*Gold Coast City Council v Halcyon Waters Community Pty Ltd* [2011] QLAC 3).

In determining the compensation for severance, market value would be the preferred basis, relying on comparable sales relevant to the land before and after the severance, using the before and after method (see Section 2.5.1, Chapter 2). However, in the absence of comparable sales, a cost approach may be adopted to determine the compensation.

In *Lenz Nominees Pty Ltd v Commissioner of Main Roads* [2012] WASC 6, which concerned compensation for the loss of 501 pine trees located on the acquired land, the Court held that the only appropriate method of compensation was the cost of replanting the trees on the retained land under the head of severance (Hyam, 2014).

4.4.2 Compensation for severance – practice

Following Jacobs (2015), the relevant inquiry is whether or not there is diminution in the value of the remaining land caused by the resumption, depending on the specific facts of the case.

As Appendix 1 shows, the paddock is severed by the acquisition, with Mickey holding two parcels after the acquisition, being the northern paddock triangular parcel of 600 sqm and the southern paddock triangular parcel (including the homestead) of 2,000 sqm. However, the bore which provides the water source for the whole paddock is on the northern parcel such that, following acquisition, the southern parcel has no access to the bore for water supply.

While Mickey may develop the retained land for low density residential use in the medium term, the preferred use for the immediate future is as a paddock. The impact of the severance is, therefore, that the retained southern parcel has no water supply such that the value of the retained southern parcel might be expected to be diminished relative to the value of land with a water supply.

4.4.2.1 Facilitating statute

As the subject property is in New South Wales, the relevant facilitating statute is the *Land Acquisition (Just Terms Compensation) Act 1991*, which states:

55 Relevant matters to be considered in determining amount of compensation

In determining the amount of compensation to which a person is entitled, regard must be had to the following matters only (as assessed in accordance with this Division) -

(c) any loss attributable to severance,

58 Loss attributable to severance

In this Act –

loss attributable to severance of land means the amount of any reduction in the market value of any other land of the person entitled to compensation which is caused by that other land being severed from other land of that person.

4.4.2.2 Compensation for severance

To determine the reduction in market value resulting from the severance of the paddock, it would be preferable to use the comparative method of valuation with the before and after method (see Section 2.5.1, Chapter 2).

While your accumulation, analysis, adjustment and application of comparable sales for agricultural land with a water supply suitable for grazing indicate a market rate of $100 psm, you are unable to find any comparable sales for agricultural land without a water supply suitable for grazing.

In the absence of comparable sales without a water supply, other methods of valuation may be considered such as:

- rounding down the rate of $100 psm based on professional experience and judgment; or
- making a percentage reduction in the value of the land based on $100psm using professional experience and judgment; or
- making a fixed capital reduction in the value of the land based on $100psm using professional experience and judgment; or
- a combination of the above.

Applying your professional experience and judgement, in the absence of comparable sales, you consider rounding down the rate of $100 psm for the value of the retained 2,000 sqm parcel of land with a water supply to $87.50 psm for the value of land without a water supply to be appropriate, being compensation for severance of $25,000, noting that this is equivalent to a reduction of 12.5% or a fixed capital reduction of $25,000 (approximately equating to the cost of sinking a bore on the retained land), which you consider to be supportive.

However, none of these methods of valuation can be supported by market evidence and so may be subject to challenge, with the valuer for the dispossessed and the valuer for the acquiring authority each potentially having equally valid but different opinions on the quantum of adjustment.

An alternative approach would be to consider claiming the cost of supplying water to the southern paddock parcel as a disturbance claim, subject to the requirement to mitigate the cost through a cost-effective means of supply. As the cost of piping water under the Happytown Freeway may be prohibitive, a cost-effective solution may be to sink a new bore on the southern paddock parcel. However, as Mickey is receiving market value compensation for the land for a purpose other than that for which it is currently used, a disturbance claim is prohibited under s61 of the *Land Acquisition (Just Terms) Compensation Act* 1991.

The compensation for severance is summarised in Table 4.4.

Table 4.4 Summary – severance compensation

Dispossessed	*Interest in land*	*Severance*
Mickey	2,000 sqm parcel of land	$25,000
Total		**$25,000**

Source: Author

4.5 Compensation for injurious affection or betterment

While the terminology in various court decisions and legal textbooks varies, adverse effects are generally referred to as injurious affection or worsement and beneficial effects are generally referred to as betterment or enhancement. For the purposes of this text, the terms injurious affection and betterment will be adopted.

Compensation for injurious affection or betterment arises where land which adjoins (with adjoins including land separated from the acquired land by a public road) or which has been severed from the acquired land is either adversely affected (injured or worsened) or beneficially affected (bettered or enhanced) by the acquisition of the acquired land by the acquiring authority.

Any compensation for injurious affection or betterment should be added to or deducted from any compensation for market value, special value or severance but does not offset disturbance (*Tolson v Roads and Maritime Services* [2014] NSWCA 161, *Mir Bros Unit Constructions Pty Ltd v Roads and Traffic Authority of NSW* [2004] NSWLEC 612). Where betterment exceeds the other value heads of compensation, then nil compensation is payable (*Brell v Penrith City Council* (1965) 11 LGRA 156).

Common examples of injurious affection and betterment include:

- the acquisition of land for a sewerage farm which, through odour, stigma and outlook, adversely affects (or injures or worsens) the remaining portion of the parent parcel or
- the acquisition of land from shop owners for a car park adjacent to the shop which benefits (or betters or enhances) the remaining land upon which the shop is located.

Without the acquisition of land injuring or bettering the parent parcel, there can be no grounds for compensation for injurious affection or betterment.

4.5.1 Compensation for injurious affection and betterment – principles

Jacobs (2015) identifies only three facilitating statutes that mention the term injurious affection and none define the term:

Queensland: *Acquisition of Land Act* 1967
Section 20(1)(a)(ii) provides for compensation for the damage caused *by the constructing authority otherwise injuriously affecting the claimant's other land.*

South Australia: *Land Acquisition Act* 1969
Section 25(1)(b)(ii) provides for compensation for *loss occasioned by reason of . . . injurious affection.*

Tasmania: *Land Acquisition Act* 1993
Sections 7H and 27(1)(e) provides for compensation where *other land belonging to the claimant is injuriously affected by the carrying out of, or the proposal to carry out, the authorised purpose.*

with the following containing associated provisions:

New South Wales: *Land Acquisition (Just Terms Compensation) Act* 1991

Section 55(f) provides compensation shall have regard to *any increase or decrease in the value of any other land of the person at the date of acquisition which adjoins or is severed from the acquired land by reason of the carrying out of, or the proposal to carry out, the public purpose for which the land was acquired.*

Western Australia: *Land Administration Act* 1997
Section 241(7) provides for compensation for a person who is also a holder of adjoining land for damage due to a reduction of *the value of that adjoining land* offset by any increases in value.

Commonwealth: *Lands Acquisition Act* 1989
Section 55(2)(a)(iv) provides for *where the acquisition has the effect of severing the acquired interest, any . . . decrease in the market value of the interest still held by the person resulting from the nature of, or the carrying out of, the purpose for which the acquired interest was acquired* with section 93 providing an over-ride for the Court to ensure that acquisition is on just terms.

while the following make no reference:

Victoria: *Land Acquisition and Compensation Act* 1986
No provision for injurious affection.

Australian Capital Territory: *Lands Acquisition Act* 1994
No provision for injurious affection.

Northern Territory: *Land Acquisition Act* 1978
No provision for injurious affection although wide discretion is provided to the Tribunal determining compensation.

The courts have provided guidance concerning the principles of injurious affection, with three key points arising from *Sisters of Charity of Rockingham v R* [1922] 2 AC 315:

- injurious affection may arise where land is severed;
- injurious affection relates to works undertaken on the part taken and their depreciating effect on the part retained and
- injurious affection relates to anticipated or future works lowering the value of retained land and not to their actual depreciating effect on completion or during construction,

being modified in *Marshall v Director-General, Department of Transport [2001] HCA 37* with a generous approach to be taken to the issue of injurious affection only applying to works undertaken on the part taken (Newton and Conolly, 2017, pages 186–187; Jacobs, 2015, page 275).

In *Commonwealth v Morison* (1972) 127 CLR 32, the Court held:

> The better view is that the loss due to injurious affection must result directly from the nature of, or the carrying out of, the purpose for which the interest was acquired.

4.5.2 Before and after method of valuation

Compensation for injurious affection is usually determined through the use of the before and after method (see Section 2.5.1, Chapter 2). The before and after method was described in *Lenz Nominees Pty Ltd v Commissioner of Main Roads* [2012] WASC 6 at 273–5 as follows:

> . . . assessment of the reduction in value of the Remaining Land must be determined by comparing the following two matters:
>
> (1) the value of the Remaining Land, at its highest and best use, before the resumption of the Taken Land without taking into account the public works; and
> (2) the value of the Remaining Land, at its highest and best use, after the resumption of the Taken Land and taking into account the public works.
>
> (Hyam, 2014, page 463)

The before and after method has been accepted by the courts in numerous cases (see, for example, *Roads and Traffic Authority of NSW v Damjanovic (2006) 146 LGERA 403*).

In applying the before and after method, the valuer will:

- value the entire property before the compulsory acquisition of part of the land, using the relevant valuation method with full supporting evidence, disregarding the public purpose and
- value the retained portion of the property after the compulsory acquisition of part of the land, using the relevant valuation method with full supporting evidence and having regard to the public purpose,

with the difference between the valuations being the compensation amount.

It should be noted that this may be positive or negative. If negative (where value after is less than before), then compensation for injurious affection will arise. If positive (where value after is greater than before – such as where smaller blocks of land sell at a higher rate than larger blocks of land), then betterment will arise and should not be double counted or should be deducted, if necessary, from the other value heads of compensation payable to the dispossessed.

4.5.3 Compensation for injurious affection – practice

As Appendix 1 shows, following the acquisition Mickey retains two parcels of land comprising a 600 sqm triangle of land in the north-western corner of the original block and a 2,000 sqm triangle of land including the homestead in the south-eastern corner of the original block.

While it is not apparent how the acquisition will cause betterment to the retained parcels, injurious affection could arise through:

- adverse impact of the freeway on the homestead including noise, dust, lack of privacy and diminution in status as the homestead is no longer surrounded by a large area of paddock and
- those parcels retained by Mickey no longer meeting Council's land size criteria for medium density residential development and only meeting Council's criteria for low density residential development resulting in significantly less development potential.

4.5.3.1 Facilitating statute

As the subject property is in New South Wales, the relevant facilitating statute is the *Land Acquisition (Just Terms Compensation) Act 1991*, which states:

55 Relevant matters to be considered in determining amount of compensation
In determining the amount of compensation to which a person is entitled, regard must be had to the following matters only (as assessed in accordance with this Division) -

(f) any increase or decrease in the value of any other land of the person at the date of acquisition which adjoins or is severed from the acquired land by reason of the carrying out of, or the proposal to carry out, the public purpose for which the land was acquired.

4.5.3.2 Compensation for injurious affection

To determine the reduction in market value resulting from the injurious affection, the preferred approach would be to use the comparative method of valuation with the before and after method (see Section 2.5.1, Chapter 2).

4.5.3.2.1 HOMESTEAD – INJURIOUS AFFECTION

The preferred approach to valuation would be to find comparable sales of similar homesteads surrounded by a large area of land unaffected by noise, dust and lack of privacy and comparable sales of similar homesteads not surrounded by a large area of land and affected by noise, dust and lack of privacy. Following the process of accumulation, analysis, adjustment and application, a valuation of the homestead before and after could be undertaken, with the difference forming the compensation for injurious affection.

However, such sales may be few and limited in comparability. In the absence of comparable sales, other methods of valuation may be considered, such as undertaking a valuation on the before basis using the accumulation, analysis, adjustment and application approach to the comparative method of valuation and then making a percentage reduction in the value of the homestead using professional experience and judgement.

While such judgement may be informed by discussions with local real estate agents and land brokers, it cannot be supported by market evidence and so may be subject to challenge, with the valuer for the dispossessed and the valuer for the acquiring authority each potentially having equally valid but different opinions on the quantum of adjustment.

Negotiations between the dispossessed and the acquiring authority may result in some works being undertaken by the acquiring authority, such as tree and bush planting, double glazing of windows and so forth, in order to mitigate the impact on the homestead. However, given the usual architectural style of country homesteads, it is likely to still be adversely affected by the freeway.

For the purposes of this example, it is assumed that the accumulation, analysis, adjustment and application of comparable sales to the homestead in the before scenario result in a determination of value of $500,000 and the valuer's judgement, informed by discussions with local market participants, indicates a reduction of 20% to reflect the impact of reduced curtilage, noise, dust and lack of privacy.

This provides an assessment of compensation for Mickey for injurious affection to the homestead of $100,000.

4.5.3.2.2 LAND PARCELS – INJURIOUS AFFECTION

Mickey retains two parcels of land comprising a 600 sqm triangle of land in the north-western corner of the original block and a 2,000 sqm triangle of land including the homestead in the south-eastern corner of the original block.

However, those parcels retained by Mickey no longer meet Council's land size criteria for medium density residential development and only meet Council's criteria for low density residential development, resulting in significantly less development potential. The injurious affection suffered by Mickey is, therefore, the difference in value between the retained land zoned for medium density development and the retained land zoned for low density development.

Your accumulation, analysis, adjustment and application of comparable sales for medium density residential development land indicate a market rate of $500 psm.

Your accumulation, analysis, adjustment and application of comparable sales for low density residential development land indicate a market rate of $250 psm.

Adopting the before and after method:

Before		
600 sqm × $500 psm	$300,000	
2,000 sqm × $500 psm	$1,000,000	$1,300,000
After		
600 sqm × $250 psm	$150,000	
2,000 sqm × $250 psm	$500,000	$650,000
Difference	$650,000	

This provides an assessment of compensation for Mickey for injurious affection to the retained land of $650,000.

4.5.3.2.3 MICKEY – COMPENSATION FOR INJURIOUS AFFECTION

The assessment of compensation for injurious affection for Mickey may be summarised as follows:

Homestead	$100,000
Retained land	$650,000
Total	$750,000

4.6 Summary and conclusions

Focusing on the key issues from a valuer's perspective, Chapters 1 and 2 introduced the concepts and principles of statutory valuation, with statutory valuation for acquisition considered in Chapters 3, 4, 5 and 6. Statutory valuation for rating and taxing is considered in Chapters 7 and 8, with regulatory valuation then considered in Chapters 9 and 10.

Chapter 1 introduced the principles of value in statutory valuation, including the concepts of market value, liberal/conservative estimates of value, highest and best use, potentiality and confirmation of a foresight. This provided the essential foundation principles of statutory valuation that are then applied in the following chapters on statutory valuation.

Chapter 2 considered the practice of statutory valuation, including valuation methods, with an emphasis on the comparative method of valuation, together with statutory valuation conventions such as the before and after method and the piecemeal method. This provided an understanding of the valuation methods and conventions that may be preferred for application in the following chapters on statutory valuation.

Statutory valuation for acquisition is considered in Chapters 3, 4, 5 and 6. Chapter 3 introduced the principles of statutory valuation for acquisition, including the statutory framework and process, with a focus on the concept of disregard of the scheme in compensation assessment and the distinction between concepts of value and worth. This provided an insight into concepts of fairness and equity in statutory valuation in the context of the disregard of compulsion.

This chapter explored the market value bases of compensation in acquisition for which statutory valuation may be required, including market value of the interest and the related market value compensation concepts of severance and injurious affection. Compensation for the market value of the interest acquired often forms the largest proportion of the compensation received by the dispossessed. While variously described in the facilitating statutes, market value usually follows a normative definition, provides no compensation for worth to the dispossessed and disregards the public purpose. The adoption of cost approaches to valuation, rather than market approaches, for compensation for relocation and reinstatement was considered, together with the role of market value in the assessment of compensation for the extinguishment of a business upon acquisition.

The related value heads of compensation of severance and injurious affection were considered and the importance of the valuer having regard to the exact language of the relevant facilitating statute was emphasised. The essential principles of compensation for severance requiring the land to have been severed and compensation for injurious affection requiring the retained land to be injured (either positively or negatively) were explored. While the preferred basis for assessing compensation for severance and injurious affection would be market value based on the comparative method of valuation, the challenge of a lack of comparable sales and the resultant need to consider other valuation approaches was investigated through the case study.

This chapter provided a broad understanding of the heads of compensation for acquisition on a value basis that may arise in statutory valuation. In the context of the case study adopted for Chapters 3, 4, 5 and 6, the value basis compensation due is summarised in Table 4.5.

Table 4.5 Summary – value basis compensation

Dispossessed	*Interest in land*	*Market value*	*Severance*	*Injurious Aff*
Mickey	Freehold – house A	$240,000		
	Freehold – house B	$250,000		
	Freehold – shop	$175,000		
	Freehold – meadow	$375,000		
	Billboard structure	$50,000		
	Freehold – paddock	$2,625,000		
		$3,715,000	$25,000	$750,000
Minnie	Leasehold – house A	$4,400		
Daisy	Leasehold (periodic tenancy) – house B	$0		
Pluto	Leasehold – shop	$5,500		
Goofy	Leasehold (licence) – meadow	$0		
Donald	Leasehold (tenancy at will) – billboard	$0		
Total		**$3,724,900**	**$25,000**	**$750,000**

Source: Author

Chapter 5 considers non-market value forms of compensation, including special value, which is noted to rarely arise, payment for solace arising from loss of home which is heavily dependent on the exact wording of the facilitating statute and the concept of disturbance for costs and losses suffered as a result of the acquisition. This provides a broad understanding of the heads of compensation for acquisition that are not necessarily on a value basis but that may arise in statutory valuation.

Chapter 6 concludes the chapters considering statutory valuation for acquisition by examining compensation for other interests such as easements and mortgages, mineral rights and profits à prendre, strata title and substrata, businesses and the emerging statutory valuation areas of native title and indigenous cultural loss. This provides an appreciation of the diversity of interests that may be subject to acquisition and so may require valuation for the assessment of compensation.

Statutory valuation for rating and taxing is considered in Chapters 7 and 8. Chapter 7 introduces the principles of land as the basis for taxation through land tax, stamp duty and rates, considering the general principles of the valuation of land for taxation using the New South Wales statutory framework as an example. This provides a general understanding of the often arcane and artificial basis of statutory valuation upon which government collects very significant amounts of taxation revenue.

Chapter 8 explores the practice of statutory valuation for rating and taxing through exemptions, concessions and allowances that have evolved over the last century, together with contemporary issues such as mass appraisal, automated valuation models and artificial intelligence, the impact of major disasters on valuation for rating and taxing and the litigation process, including the valuer as an expert witness. This provides an insight into the continued evolution of statutory valuation to address issues as they arise over time, highlighting the importance of a principles-based approach to statutory valuation.

Regulatory valuation, considered in the context of the preparation of valuations for inclusion in financial reporting, is considered in Chapters 9 and 10. Chapter 9 explores the principles of regulatory valuation with regard to the inter-related roles of international accounting standards, international financial reporting standards and international valuation standards, focusing on the primacy of fair value in international financial reporting standards. This provides an understanding of the fundamental conceptual differences and bases between statutory valuation and regulatory valuation.

Finally, Chapter 10 concludes regulatory valuation by exploring the practice of valuation approaches and methods relevant for regulatory valuation such as the income approach and the cost approach and the various valuation methods there within. This provides an appreciation of the wider range of valuation approaches and methods available to the valuer for regulatory valuation, reflecting that, unlike statutory valuation, there is no requirement for defence in Court (negligence claims excepted).

References

Brown, D (1991) *Land acquisition*, Lexis Nexis Butterworths, Chatswood
Brown, D (2009) *Land acquisition*, Lexis Nexis Butterworths, Chatswood
Hyam, A (2014) *The law affecting valuation of land in Australia*, The Federation Press, Annandale
Jacobs, M (2015) *Law of compulsory land acquisition*, Thomson Reuters, Pyrmont
Newton, G and Conolly, C (2017) *Land acquisition*, Lexis Nexis Butterworths, Chatswood
Parker, D (Ed) (2022) *Principles and practice of property valuation in Australia*, Routledge, Abingdon
Parker, D (2024) *Introduction to property valuation in Australia*, Routledge, Abingdon

Appendix 1 – Case study

A1.1 Overview

Mickey runs a large horse breeding and racing business, based in Happytown, a small country town in NSW, comprising a 10,400 sqm block of land located on the edge of Happytown together with several hundred hectares of grazing land surrounding Happytown.

Prior to the compulsory acquisition, the 10,400 sqm block of land owned by Mickey was surrounded by Mouse Street, Duck Street and Hound Street, as shown in the before graphic (2.0). The block comprised a paddock to which horses would be brought from Mickey's surrounding grazing lands for grazing, shoeing, veterinary attention and general care, together with a homestead and a water bore, a meadow used for grazing horses with a billboard, a blacksmith's shop and two houses, each of which is separately fenced off, with the entire 10,400 sqm being zoned for medium density residential development.

The proposed acquisition will facilitate the construction, operation and maintenance of the Happytown Freeway through Mickey's block of land, as shown in the after graphic (3.0). The land to be acquired is marked by an arrow in the graphic below. The acquisition will be gazetted on 31 December 2024.

As a result of the compulsory acquisition, the portions of the block retained by Mickey north and south of the proposed Happytown Freeway will no longer meet Council's land size criteria for medium density residential development but will meet Council's criteria for low density residential development.

A1.2 Before graphic

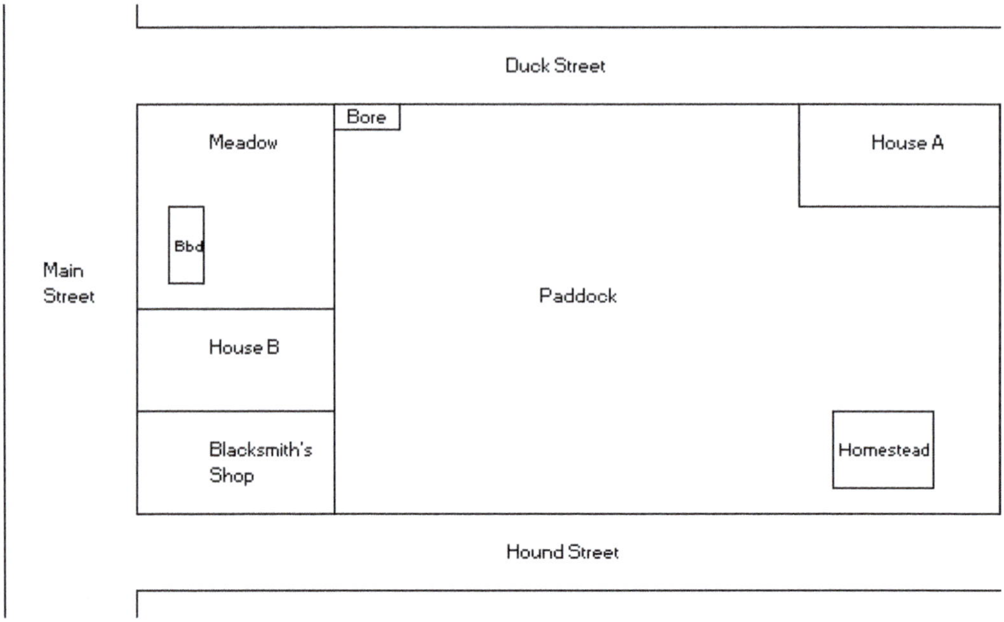

A1.3 After graphic

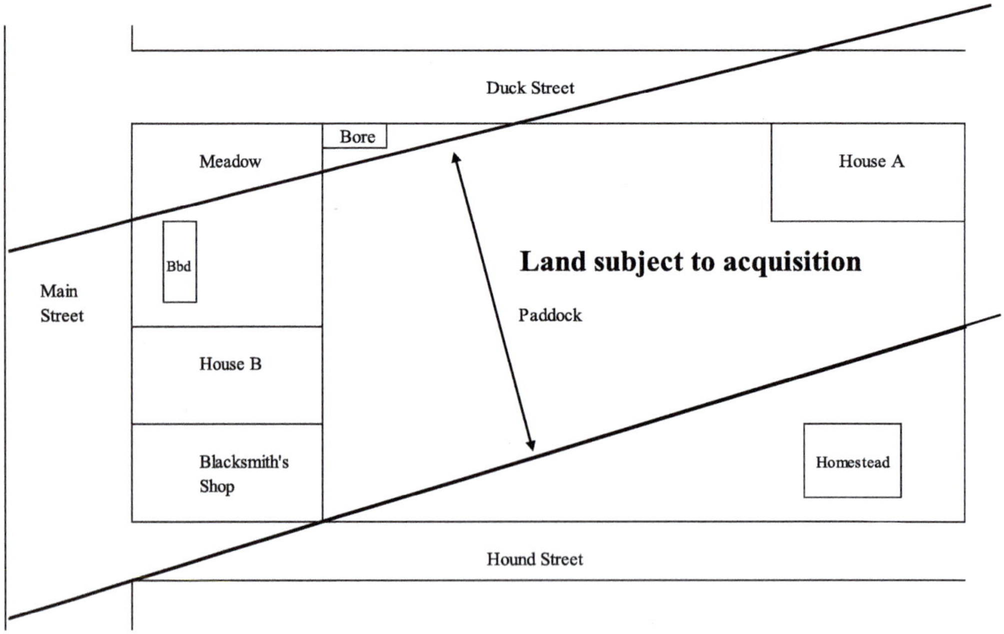

A1.4 Land areas

The respective land areas are as follows:

Area of Mickey's block	10,400 sqm
Area of homestead and paddock	7,400 sqm
Area of shop site	600 sqm
Area of house A and B sites	600 sqm each
Area of meadow	1,200 sqm
Area to be acquired	7,800 sqm
Comprising:	
Paddock	5,250 sqm
House A	600 sqm
House B	600 sqm
Shop	600 sqm
Meadow	750 sqm
Portion of the block retained north of the proposed freeway	600 sqm (450 sqm meadow, 150 sqm paddock)
Portion of the block retained south of the proposed freeway	2,000 sqm

A1.5 Parties

Mickey occupies the homestead and the paddock.

Mickey leases house A to Minnie on a five-year lease expiring 31 December 2025 at $100 pcm. Minnie has been the occupant of house A as her principal place of residence for over a decade. Minnie is very elderly and moving house is a very traumatic event causing extreme anxiety. The acquiring authority requires possession on 31 December 2024.

Mickey leases house B to Daisy on a month-to-month tenancy at $100 pcm. Daisy has been the occupant of house B as her principal place of residence for three years. Daisy was thinking of moving to a smaller residence and welcomes the opportunity to relocate. The acquiring authority requires possession on 31 December 2024.

Mickey leases the shop to Pluto on a ten-year lease expiring 31 December 2025 at $250 pcm. Pluto runs a blacksmith's shop, with existing use rights, which services Mickey's horses in the paddock as well as minor metalwork jobs for residents of Happytown, showing future maintainable profits of $1,000 pa and holding stock worth $5,000. Pluto has decided it is time to retire and will not relocate his blacksmith's shop elsewhere following the compulsory acquisition.

Mickey has given a licence to Goofy to use the meadow for six months expiring 31 December 2024 at $1 pcm.

Mickey owns the billboard and has given a tenancy at will to Donald to use the billboard at $60 pcm.

A1.6 Facilitating statute

The proposed acquisition notice stated that the land was to be acquired *for the purposes of the Roads Act . . . in connection with the construction, operation and maintenance of the Happy-town Freeway.*

A1.7 Valuation information

Your accumulation, analysis, adjustment and application of comparable sale transactions for houses A and B indicate that the market value for each house and curtilage is $250,000.

Your accumulation, analysis, adjustment and application of comparable rental transactions for houses A and B indicate that the market rental for each house is $500 pcm.

Your accumulation, analysis, adjustment and application of comparable sale transactions for short term residential leasehold interests indicate that such interests trade on a capitalisation rate of 10%.

Your accumulation, analysis, adjustment and application of comparable rental transactions for the blacksmith's shop indicate that the market rental for the shop is $750 pcm.

Your accumulation, analysis, adjustment and application of comparable sales for freehold shops and curtilage indicate that shops trade on a capitalisation rate of 5%.

Your accumulation, analysis, adjustment and application of comparable sale transactions for short term leasehold shop interests indicate that such interests trade on a capitalisation rate of 10%.

Your analysis of the turnover of the blacksmith's shop shows future maintainable profits of $1,000 pa.

Your accumulation, analysis, adjustment and application of comparable business sales show that blacksmith's businesses trade on a capitalisation rate of 10%.

Your accumulation, analysis, adjustment and application of comparable rentals for meadow grazing indicate a market rental for 1200 sqm to be $120 pcm.

Your accumulation, analysis, adjustment and application of comparable rentals for billboards indicate a market rental to be $120 pcm.

Your accumulation, analysis, adjustment and application of comparable sales for billboard businesses indicate that billboard businesses trade on a capitalisation rate of 10%.

Your analysis of the turnover of the horse breeding and training business conducted on the land shows future maintainable profits of $10,000 pa.

Your accumulation, analysis, adjustment and application of comparable business sales show that horse breeding and training businesses trade on a capitalisation rate of 15%.

Your accumulation, analysis, adjustment and application of comparable sales for pastoral land with a water supply suitable for grazing indicate a market rate of $100 psm.

Your accumulation, analysis, adjustment and application of comparable sales for low density residential development land indicate a market rate of $250 psm.

Your accumulation, analysis, adjustment and application of comparable sales for medium density residential development land indicate a market rate of $500 psm.

You have received advice that the replacement cost of the billboard is $250,000 and that the existing billboard has a potential remaining usable life of ten years, having been erected 40 years ago.

A1.8 Disturbance claims

Mickey has submitted four invoices for costs associated with the compulsory acquisition, being \$10,000 for legal advice on the acquisition, \$5,000 for legal advice on likely success in an action in the Land and Environment Court, \$5,000 for valuer's fees and \$13,838 for stamp duty on two new houses.

Minnie has submitted three invoices for costs associated with vacating house A, being \$100 for a solicitor to review the lease of her new rental house, \$500 for removalists and \$1,000 for her friend who helped Minnie find a new rental house to move into.

Daisy has submitted three invoices for costs associated with vacating house B, being \$100 for a solicitor to review the lease of her new rental house, \$500 for removalists and \$2,500 for the cost of repainting the interior of her new rental house.

Pluto has submitted four invoices for costs associated with vacating the blacksmith's shop, being \$500 for the hire of a van to move stock from the shop to his home garage, \$150 for an advice of retirement sign, \$1,000 for losses suffered in a closing down sale and \$500 for setting up his business landline, fax machine and internet connection in his home.

Donald has submitted one invoice for \$10,000 for loss of profits from billboard advertising for a period of 12 months after the date of acquisition.

Goofy has submitted one invoice for the costs associated with moving his horse from the meadow to another meadow, which is within walking distance, being \$50,000.

5 Compulsory acquisition

Compensation for special value, solatium and disturbance

5.1 Introduction

This book is an introduction to two very different valuation worlds in Australia – the complex and often arcane national realm of statutory valuation and the dynamic global empire of regulatory valuation. Statutory valuation is different from regulatory valuation, with the former being based on relevant applicable statutes and being capable of challenge on appeal in Court while the latter is based on relevant applicable regulations and incapable of challenge in Court (negligence claims excepted).

As an introductory textbook on statutory and regulatory valuation, this book is a companion to the Australian valuation textbooks, *Introduction to Property Valuation in Australia* (Parker, 2024) and *Principles and Practice of Property Valuation in Australia* (Parker, 2022), which provide a wider analysis of key principles underlying property valuation and current techniques and issues in the practice of property valuation for major sectors of the Australian property market.

Statutory valuation in Australia is most commonly encountered in valuations for rating, taxing and acquisition but may arise wherever a statute prescribes a basis of or approach to valuation. Regulatory valuation is most commonly encountered in valuations for financial statements but may arise wherever a regulation prescribes a basis of or approach to valuation.

Rating and taxing of land raise significant amounts of revenue for State Governments and Councils in Australia, based on statutory valuation. For example, the NSW 2025–26 Budget Statement estimates land tax revenue for 2024–2025 to be $8.2 billion with the City of Brisbane Annual Plan and Budget 2024–2025 estimating proposed rates and utility charges to be $1.51 billion.

Acquisition of land, property and businesses commonly arises for major infrastructure projects and may be by agreement or by compulsion, based on statutory valuation. For example, such major infrastructure projects in Australia currently include the multi-billion-dollar Sydney Metro, Melbourne Suburban Rail Loop and Melbourne to Brisbane Inland Rail projects.

Valuation of land and property holdings based on regulatory valuation arises for all reporting entities in Australia that publish financial statements upon which it is reasonable to expect that users will rely for decision making. For example, at June 2025 there are over 2,000 reporting entities listed on the Australian Stock Exchange which each publish financial statements, with the financial statements for Goodman Group, as at 30 June 2024, showing the fair value of total property assets to be $20.2 billion, based on regulatory valuation.

Focusing on the key issues from a valuer's perspective, the concepts and principles of statutory valuation are introduced in Chapters 1 and 2, with statutory valuation for acquisition considered in Chapters 3, 4, 5 and 6 and statutory valuation for rating and taxing considered in Chapters 7 and 8. Regulatory valuation is then considered in Chapters 9 and 10.

DOI: 10.1201/9781032616957-5

Chapter 1 introduced the principles of value in statutory valuation, including the concepts of market value, liberal/conservative estimates of value, highest and best use, potentiality and confirmation of a foresight. This provided the essential foundation principles of statutory valuation that are then applied in the following chapters on statutory valuation.

Chapter 2 considered the practice of statutory valuation including valuation methods for statutory valuation with an emphasis on the comparative method of valuation, together with statutory valuation conventions such as the before and after method and the piecemeal method. This provided an understanding of the valuation methods and conventions that may be preferred in the following chapters on statutory valuation.

Statutory valuation for land acquisition is considered in Chapters 3, 4, 5 and 6, supported by an evolving case study in Appendix 1 of each chapter, including the principal bases of compensation being:

- compensation for market value for the interest acquired;
- compensation for severance;
- compensation for injurious affection
- compensation for special value;
- compensation for solatium or disadvantage resulting from relocation; and
- compensation for disturbance.

Chapter 3 introduced the principles of statutory valuation for land acquisition, including the statutory framework and process, with a focus on the concept of the disregard of the scheme in compensation assessment and the distinction between concepts of value and worth. This provided an insight into concepts of fairness and equity in statutory valuation in the context of the requirement to disregard the compulsory nature of the acquisition.

Chapter 4 explored the market value bases of compensation in acquisition for which statutory valuation may be required, including market value of the interest and the related market value compensation concepts of severance and injurious affection. This provided a broad understanding of the heads of compensation for acquisition on a value basis that may arise in statutory valuation.

This chapter considers non-market value forms of compensation, including special value, which is noted to rarely arise, payment for solace arising from loss of the primary place of residence and the concept of disturbance for costs and losses suffered as a result of the acquisition. This provides a broad understanding of the heads of compensation for acquisition that are not, necessarily, on a value basis but which may arise in statutory valuation.

Chapter 6 concludes the chapters considering statutory valuation for acquisition by examining compensation for other interests such as easements and mortgages, mineral rights and profits à prendre, strata title and substrata, businesses and the emerging statutory valuation areas of native title and indigenous cultural loss. This provides an appreciation of the diversity of interests that may be subject to acquisition and so may require valuation for the assessment of compensation.

Statutory valuation for rating and taxing is considered in Chapters 7 and 8. Chapter 7 introduces the principles of land as the basis for taxation through land tax, stamp duty and rates, considering the general principles of the valuation of land for taxation using the New South Wales statutory framework as an example. This provides a general understanding of the often arcane and artificial basis of statutory valuation upon which Government collects very significant amounts of taxation revenue.

Chapter 8 explores the practice of statutory valuation for rating and taxing through exemptions, concessions and allowances that have evolved over the last century, together with contemporary issues such as mass appraisal, automated valuation models and artificial intelligence, the impact of major disasters on valuation for rating and taxing and the litigation process including the valuer as an expert witness. This provides an insight into the continued evolution of statutory valuation to address issues as they arise over time, highlighting the importance of a principles-based approach to statutory valuation.

Regulatory valuation, considered in the context of the preparation of valuations for inclusion in financial reporting, is considered in Chapters 9 and 10. Chapter 9 explores the principles of regulatory valuation with regard to the inter-related roles of international accounting standards, international financial reporting standards and international valuation standards, focusing on the primacy of fair value in international financial reporting standards. This provides an understanding of the fundamental conceptual differences and bases between statutory valuation and regulatory valuation.

Finally, Chapter 10 concludes regulatory valuation by exploring the practice of valuation approaches and methods relevant for regulatory valuation such as the income approach and the cost approach and the various valuation methods there within. This provides an appreciation of the wider range of valuation approaches and methods available to the valuer for regulatory valuation reflecting that, unlike statutory valuation, there is no requirement for defence in Court (negligence claims excepted).

Accordingly, this chapter will now consider:

- compensation for special value;
- compensation for solatium (or disadvantage resulting from relocation);
- principles of compensation for disturbance and
- practice of compensation for disturbance, using the evolving case study in Appendix 1.

5.2 Compensation for special value

This section seeks to guide the valuer through the valuation aspects of the key issues concerning compensation for special value from a valuer's viewpoint. Greater detail on the legal aspects of the key issues may be found in Jacobs (2015), Hyam (2014), Brown (2009) and Newton and Conolly (2017).

5.2.1 Compensation for special value – principles

Section 3.3.1.1 in Chapter 3 considered special value, noting that the historic ability of a Judge to award compensation for special value contributed to the sense of fairness in compensation with acquisition then being considered to be on just terms.

Prior to the introduction of the various facilitating statutes, the Courts adopted the concept of Judge determined compensation for *value in use* which was a combination of market value, special value, severance and disturbance with special value being something other than market value.

Therefore, prior to the introduction of the various facilitating statutes, a body of case law developed which is informative when compared to the very specific provisions of the various facilitating statutes currently applying in Australia.

5.2.1.1 Pre-facilitating statutes period

The concept of special value was established in *Pastoral Finance Association Ltd v Minister* [1914] AC 1083 where the Court held that the dispossessed was entitled to more than market value, being entitled to compensation based on the value of the land to them:

> Probably the most practical form in which the matter can be put is that they were entitled to that which a prudent man in their position would have been willing to give for the land sooner than fail to obtain it.
>
> (1088)

Special value arises out of the use to which the owner has, or will shortly, put the land, arising out of a peculiar feature of the land (a positive advantage or an exemption from disadvantage) which would extract a higher offer from a potential purchaser than the market value of the land – special value has to be something more than simply market value (Newton and Conolly, 2017).

Therefore, special value may arise in the dispossession of a monumental stonemason's yard, adjacent a cemetery, where the peculiar use of the land gives rise to a scenario where the value is greater than the market value for the land.

The oft quoted example of grounds for special value is that of a blacksmith's forge adjacent a racecourse, cited by Callinan J in *Boland v Yates Property Corp Pty Ltd* (1999) 74 ALJR 209 at 269:

> There will in practice be few cases in which a property does have a special value for a particular owner. Obviously neither sentiment nor a long attachment to it will suffice. The special quality must be a quality that has an economic significance to the owner. A possible case would be one in which, for example, a blacksmith operates a forge in the vicinity of a racetrack on land zoned for residential purposes as a protected non-conforming use, the right to which might be lost on a transfer of ownership or on interruption of the protected use. Such a property will have a special value for its blacksmith owner, and perhaps another blacksmith who might be able to comply with the relevant requirements to enable him to continue the use but no one else.

Significantly, special value in case law is limited to an objective assessment of economic or commercial value and does not include subjective assessments of sentimental or emotional value or any function of the skill or expertise of the owner (Newton and Conolly, 2017).

Valuers should start from the position that instances of special value (or their facilitating statute equivalent) are very rare and usually require a specific set of circumstances to be proven to exist in a specific matter. However, it may be contended that the historic case law principles of special value potentially provided the dispossessed with a greater opportunity for compensation than will arise today from a very literal interpretation of the special value equivalents in current facilitating statutes.

5.2.1.2 Post-facilitating statutes period

Following the introduction of the various facilitating statutes, that which may be compensated as special value was codified in the statute and Judges limited to awarding only compensation permitted by a literal interpretation of the facilitating statute. Therefore, valuers should focus very closely on the exact wording of the facilitating statute and ensure that it is fulfilled if attempting to ground a compensation claim for special value.

Special value is variously defined in the facilitating statutes as follows (Jacobs, 2015):

Commonwealth: *Lands Acquisition Act* 1989
Section 55(2)(a)(ii):

> The value, on the day of acquisition, of any financial advantage, additional to market value, to the person incidental to the person's ownership of the interest.

New South Wales: *Land Acquisition (Just Terms Compensation) Act* 1991
Section 55:

> In determining the amount of compensation to which a person in entitled, regard must be had to the following matters only (as assessed in accordance with this Division):
>
> (b) any special value of the land to the person on the date of its acquisition,

Section 57:
In this Act:

> "special value" of land means the financial value of any advantage, in addition to market value, to the person entitled to compensation which is incidental to the person's use of the land.

Victoria: *Land Acquisition and Compensation Act* 1986
Section 40:

> The definition of "special value" in relation to an interest in land means the value of any pecuniary advantage, in addition to market value, to a claimant which is incidental to his ownership or occupation of that land.

Section 41(1)(b):

> Except as otherwise provided in this Part, in assessing the amount of compensation payable to a claimant in respect of an interest in land which is acquired under this Act, regard must be had to the following factors:
>
> (b) any special value of the land to the person on the date of its acquisition;

Queensland: *Acquisition of Land Act* 1967
Section 20 (2):

> Compensation shall be assessed according to the value of the estate or interest of the claimant in the land taken on the date when it was taken.

Western Australia: *Land Administration Act* 1997

Section 241(6)(e):

(6) Regard is to be had to the loss or damage, if any, sustained by the claimant by reason of –

(e) any other facts which the acquiring authority, the court, or the State Administrative Tribunal considers it just to take into account in the circumstances of the case.

South Australia: *Land Acquisition Act* 1969
Section 25(1)(e):

[T]he compensation payable under this Act in respect of the acquisition of land shall be determined according to the following principles:

. . .

any special suitability or adaptability of the land for any purpose shall not be taken into account if it could be applied to that purpose in pursuance only of statute, or if the suitability or adaptability is peculiar to the purposes or requirements of a particular person or of any Government or local governing authority but any bona fide offer to acquire the land made before the passing of the special Act shall be taken into account.

Tasmania: *Land Acquisition Act* 1993
Section 27(1)(b)(i) and (ii):

(1) In determining compensation under this Act, regard is to be had to the following matters:

(b) any special value the estate in the subject land may have to the claimant which is –

(i) a financial advantage incidental to the claimant's ownership of that estate; and
(ii) in addition to its market value;

Northern Territory: *Land Acquisition Act* 1978
Section 66 as read with Sch 2, r 2(b):

[T]he value of any additional advantage to the claimant incidental to his ownership, or occupation of, the acquired land.

Australian Capital Territory: *Lands Acquisition Act* 1994
Section 45(1) and (2)(a)(ii):

(1) The amount of compensation to which a person is entitled under this part in respect of the acquisition of an interest in land is such amount as, having regard to all relevant matters, will justly compensate the person for the acquisition.
(2) In assessing the amount of compensation to which the person is entitled, regard shall be had to all relevant matters, including –

(a) except in a case to which paragraph (b) applies –

(ii) the value, on the day of the acquisition, of any financial advantage, additional to market value, to the person incidental to the person's ownership of the interest;

While the facilitating statute of the Commonwealth, New South Wales, Victoria, Tasmania, the Australian Capital Territory and the Northen Territory each directly refer to a notion of special value, those of Queensland, Western Australia and South Australia do not.

As will be noted, special value, where appearing in the facilitating statutes, echoes concepts from the pre-facilitating statutes era such as special value being in addition to market value, being of a financial nature and relating to the use of the land. As such, this definition implies that for special value to be payable, there must first be a market value for the interest acquired. It is important, however, for valuers to have careful regard to the exact text of the relevant facilitating statute prior to determining if compensation for special value is applicable.

Whether or not the pre-facilitating statutes era special value case law applies in Queensland, Western Australia and South Australia is described by Jacobs (2015) as *a difficult question in respect of which it is hoped that greater clarity will be obtained from subsequent High Court decisions.*

5.2.2 Compensation for special value – practice

In determining if compensation for special value is applicable in the case study in Appendix 1, the valuer should have regard to the relevant facilitating statute.

5.2.2.1 Facilitating statute

As the subject property is in New South Wales, the relevant facilitating statute is the *Land Acquisition (Just Terms Compensation) Act* 1991 which states:

> Section 55(b):
>
> In determining the amount of compensation to which a person in entitled, regard must be had to the following matters only (as assessed in accordance with this Division):
>
> (b) any special value of the land to the person on the date of its acquisition,
>
> Section 57:
>
> *In this Act:*
>
> "special value" of land means the financial value of any advantage, in addition to market value, to the person entitled to compensation which is incidental to the person's use of the land.

5.2.2.2 Compensation for special value

As special value under section 57 is in addition to market value, only those dispossessed interests that qualify for market value compensation should be considered by the valuer – being, in this example, Mickey, Minnie and Pluto.

To qualify for special value under section 57, there must be an *advantage* which is of *financial value* and that is *incidental to the person's use of the land.* Considering the following uses of the land by each of the dispossessed:

- Mickey – residential occupation and horse breeding and racing;
- Minnie – residential occupation; and
- Pluto – blacksmith's shop,

only Pluto blacksmith's shop enjoys an advantage, being proximity and adjacency to the paddock, that would be of financial value and is incidental to Pluto's use of the land.

Quantifying special value is extremely challenging as it is in addition to market value. Consistent with *Pastoral Finance Association Ltd v Minister* [1914] AC 1083, special value could be considered to be *that which a prudent man in their position would have been willing to give for the land sooner than fail to obtain it* (1008).

As special value is not represented in comparable sales, a valuer will need to exercise judgment and determine either a capital amount or a percentage above the market value to be special value. As there is no basis in sales evidence, two equally competent and experienced valuers could determine different capital amounts or percentages of market value to be special value.

The market value compensation for Pluto for the 12 months unexpired of his lease was determined to be $5,500 and so the valuer may determine that special value in the order of 20% is appropriate and award $1,100. As there is no basis for such assessment, other than judgment and experience, defence of this determination may be challenging.

Jacobs (2015, page 509) notes than in *T&P Holdings Ltd v Municipality of Metropolitan Toronto* [1972] 1 LCR 340 (CA, Ont) at 340, the Court stated that the determination of special value as the present-day cash value of a potentiality to be achieved in the future is such a difficult exercise that:

> . . .(a)t best, it is a guess rather than an estimate but it is a guess which the claimant is entitled to have taken in his favour, and it is a guess which can be supplied as readily by a court as an appraiser.

5.3 Compensation for solatium

Compensation for solatium is a payment for consolation or solace following the dispossession of an interest. It represents a contribution by the acquiring authority towards solace for injured feelings and the disruption, inconvenience, annoyance, distress, hardship, trauma, depression, sentimental attachment, sense of loss and adverse effect on the dispossessed's family or other unspecified loss suffered by the dispossessed as a result of the acquisition – *It is a kind of sweetener, perhaps, reflecting some kind of apology* (Newton and Conolly, 2017, page 189).

Related to the concept of solatium is the concept of compensation for the compulsory nature of the acquisition of land. Baum et al. (2014) note that in the UK, prior to the introduction of the *Acquisition of Land Act* 1919, it had become accepted practice to add 10% to the agreed compensation as consolation for the compulsory nature of the expropriation, though this is no longer allowed. Similarly, in the relevant facilitating statutes in Australia (with the exception of the Commonwealth, the Australian Capital Territory and Western Australia), there is no recognition of the compulsory nature of the acquisition in the compensation assessment.

5.3.1 Compensation for solatium – principles

Solatium is variously defined in the facilitating statutes as follows (Jacobs, 2015; Hyam, 2014):

Commonwealth: *Lands Acquisition Act* 1989

Section 61:

(1) This section applies where:

(a) an interest in land is acquired from a person by compulsory process; and
(b) the interest entitled the person to occupy a dwelling on the land; and
(c) immediately before the acquisition the person was occupying the dwelling as his or her principal place of residence; and
(d) because of the acquisition, the person has ceased to be entitled to occupy the dwelling as his or her principal place of residence.

(2) The amount of compensation to which the person is entitled in respect of the acquisition is the sum of $10,000 (or that amount as indexed by section 126) and the greater of the following amounts:

(a) the amount of compensation to which the person would, apart from this section, be entitled;
(b) the amount necessary to reimburse the person for the costs of acquiring a reasonably equivalent interest in land that entitles the person to occupation of a reasonably equivalent dwelling.

New South Wales: *Land Acquisition (Just Terms Compensation) Act* 1991

Section 55(e):

In determining the amount of compensation to which a person in entitled, regard must be had to the following matters only (as assessed in accordance with this Division):

(e) the disadvantage resulting from relocation,

Section 60:

(1) In this Act:

"disadvantage resulting from relocation" means non-financial disadvantage resulting from the necessity of the person entitled to compensation to relocate the person's principal place of residence as a result of the acquisition.

(2) The maximum amount of compensation in respect of the disadvantage resulting from relocation is $75,000. (Note Schedule 1A provides for the amendment of this section to enable the maximum amount of compensation to be increased by regulation and for the automatic indexation of the maximum amount in line with inflation.)
(3) In assessing the amount of compensation in respect of the disadvantage resulting from relocation, all relevant circumstances are to be taken into account, including:

(a) the interest in the land of the person entitled to compensation; and
(b) the length of time the person has resided on the land (and in particular whether the person is residing on the land temporarily or indefinitely); and
(c) the inconvenience likely to be suffered by the person because of his or her removal from the land; and
(d) the period after the acquisition of the land during which the person has been (or will be) allowed to remain in possession of the land.

(4) Compensation is payable in respect of the disadvantage resulting from relocation if the whole of the land is acquired or if any part of the land on which the residence is situated is acquired.
(5) Only one payment of compensation in respect of the disadvantage resulting from relocation is payable for land in separate occupation.
(6) However, if more than one family resides on the same land, a separate payment may be made in respect of each family if:

(a) the family resides in a separate dwelling house; or
(b) the Minister responsible for the authority of the State approves of the payment.

(7) If separate payments of compensation are made, the maximum amount under subsection (2) applies to each payment, and not to the total payments.

Victoria: *Land Acquisition and Compensation Act* 1986

Section 44*:*

(1) The amount of compensation may be increased by such amount, not exceeding 10% of the market value of the land, by way of solatium as is reasonable to compensate the claimant for intangible and non-pecuniary disadvantages resulting from the acquisition.
(2) In assessing the amount payable under sub-section (1), there must be taken into account all relevant circumstances applicable to the claimant including, without limiting the generality of the foregoing-

(a) the interest of the claimant in the acquired land; and
(b) the length of time during which the claimant had occupied the land; and
(c) the inconvenience likely to be suffered by the claimant by reason of removal from the land; and
(d) the period of time after the acquisition of the land during which the claimant has been, or will be, allowed to remain in possession of the land; and
(e) the period of time during which, but for the acquisition of the land, the claimant would have been likely to continue to occupy the land; and
(f) the age of the claimant; and
(g) where the claimant at the date of acquisition is occupying the land as the claimant's principal place of residence, the number, age and circumstances of other people (if any) living with the claimant.

(3) If no solatium is paid to a claimant, a person other than a claimant who, at the date of acquisition, had occupied the acquired land for a continuous period of not less than 12 months before that date as the person's principal place of residence may claim from the Authority such amount, not exceeding 10% of the market value of the land, by way of solatium as is reasonable to compensate the person for intangible and non-pecuniary disadvantages resulting from the acquisition.
(4) In determining the amount payable under sub-section (3), there must be taken into account all relevant circumstances applicable to the person, including the matters referred to in sub-section (2)(b), (c), (d), (e), (f) and (g).
(5) If the Authority rejects a claim for solatium made by a person under sub-section (3), that person is to be taken to be a claimant and the claim to be a disputed claim for the purposes of this Act.

Queensland: *Acquisition of Land Act* 1967
There is no provision for payment of solatium.

Western Australia: *Land Administration Act* 1997

Section 241(8) and (9):

(8) If the interest in land is taken without agreement, an amount considered by the court or the State Administrative Tribunal or, for the purpose of making an offer, by the acquiring authority, appropriate to compensate for the taking without agreement may be added to the award or offer.
(9) The additional amount under sub-section (8) must not be more than 10% of the amount otherwise awarded or offered, unless the court or the State Administrative Tribunal, or, for the purposes of making an offer, the acquiring authority, is satisfied that exceptional circumstances justify a higher amount.

South Australia: *Land Acquisition Act* 1969

Section 25(1)(g):

no allowance shall be made on account of the fact that the acquisition is effected without the consent, or against the will, of any person

Tasmania: *Land Acquisition Act* 1993

While there is no specific provision for payment of solatium, such payment is probably covered in the following section:

Section 30(3):

(1) In determining compensation under this Act, regard is to be had to the following matters:

(g) except as provided in this Part, such other matters as the acquiring authority, the Court or an arbitrator may consider to be relevant.

Notwithstanding that compensation is determined in accordance with sub-section (2), additional compensation may be awarded in respect of any hardship that the claimant may suffer because the claimant cannot establish himself or herself in a suitable residence solely by reason of age, infirmity or want of means.

Northern Territory: *Land Acquisition Act* 1978

Schedule 2, r 9:

(1) If the claimant, during the period commencing on the date on which the notice of proposal was served and ending on the date of acquisition:

(a) occupied the acquired land as his principal place of residence; and
(b) held an estate in fee simple, a life estate or a leasehold interest in the acquired land,

the amount of compensation otherwise payable under this Schedule may be increased by the amount which the Tribunal considers will reasonably compensate the claimant for intangible disadvantages resulting from the acquisition.

(2) In assessing the amount payable under subrule (1), the Tribunal shall have regard to:

(a) the interest of the claimant in the land;
(b) the length of time during which the claimant resided on the land;
(c) the inconvenience likely to be caused to the claimant by reason of his removal from the acquired land;
(d) the period after the acquisition of the land during which the claimant has been, or will be, allowed to remain in possession of the land;
(e) the period during which the claimant would have been likely to continue to reside on the land; and
(f) any other matter which is, in the Tribunal's opinion, relevant to the circumstances of the claimant.

Australian Capital Territory: *Lands Acquisition Act* 1994

Section 51:

(1) This section applies where:

(a) an interest in land is acquired from a person by compulsory process; and
(b) the interest entitled the person to occupy a dwelling on the land; and
(c) immediately before the acquisition the person was occupying the dwelling as his or her principal place of residence; and
(d) because of the acquisition, the person has ceased to be entitled to occupy the dwelling as his or her principal place of residence.

(2) The amount of compensation to which the person is entitled in respect of the acquisition is the sum of $15,000 (or that amount as indexed by section 105) and the greater of the following amounts:

(a) the amount of compensation to which the person would, apart from this section, be entitled;
(b) the aggregate of the costs to the person of acquiring a reasonably equivalent interest in land that entitles the person to occupation of a reasonably equivalent dwelling.

While the facilitating statute for Queensland does not make provision for solatium, the facilitating statutes for the other States and Territories provide for a range of purposes for compensation under the general heading of solatium:

- Commonwealth focuses on compulsion;
- New South Wales focuses on non-financial disadvantage due to relocation;
- Victoria focuses on intangible and non-pecuniary disadvantages;
- Western Australia focuses on compulsion or exceptional circumstances;
- South Australia focuses on no allowance for compulsion;
- Tasmania focuses on hardship; and
- Northern Territory focuses on intangible disadvantages; and
- Australian Capital Territory focuses on compulsion.

Further, while the facilitating statutes for the Commonwealth, New South Wales and the Australian Capital Territory specify an indexed maximum amount, those for Victoria and Western Australia specify a cap of 10% with the amount in Tasmania and the Northern Territory unspecified.

In all cases, the amount to be awarded as compensation for solatium is discretionary and may not be the full amount permitted, depending on the circumstances to which regard is required.

Accordingly, it is essential that the valuer read the facilitating statute very carefully in order to determine that which may or may not be compensated under the general heading of solatium. Further, it should be noted that the facilitating statute of Western Australia may permit compensation for the compulsory nature of the acquisition, with compulsion otherwise generally disregarded in compensation assessment in Australia and specifically disregarded in South Australia.

Newton and Conolly (2017) note that compensation for solatium is a discretionary matter for the Court, is not automatically awarded and may be restricted to the acquisition of residential property with evidence required to support a claim in most jurisdictions.

In *Rees v Minster for Planning and Homes* (1991) 76 LGRA 167, the Court rejected a claim for solatium on the basis that the claimants had purchased the land at a time when there was already in existence an approved reservation of it as public open space. Further, in *Rabik and Rabik v Roads and Traffic Authority (NSW)* [1997] NSWLEC 141, the Court declined to award solatium as the request to acquire came from the dispossessed such that there was no evidence of any necessity to relocate (Jacobs, 2015).

Hyam (2014) notes that in *Kontos v Roads and Traffic Authority of NSW* (1992) 75 LGRA 218, the Court declined to pay compensation for solatium as the dispossessed occupied the residence, adjoining a shop, *merely as a place to sleep during weekdays* and *could not be considered as truly the home of the applicants*.

5.3.2 Compensation for solatium – practice

In determining if compensation for solatium is applicable in the case study in Appendix 1, the valuer should have regard to the following.

5.3.2.1 Facilitating statute

As the subject property is in New South Wales, the relevant facilitating statute is the *Land Acquisition (Just Terms Compensation) Act* 1991 which states:

Section 55(e):

> In determining the amount of compensation to which a person in entitled, regard must be had to the following matters only (as assessed in accordance with this Division):
>
> (e) the disadvantage resulting from relocation,

Section 60:

(1) In this Act:

> "disadvantage resulting from relocation" means non-financial disadvantage resulting from the necessity of the person entitled to compensation to relocate the person's principal place of residence as a result of the acquisition.

(2) The maximum amount of compensation in respect of the disadvantage resulting from relocation is $75,000. (Note Schedule 1A provides for the amendment of this section to enable the maximum amount of compensation to be increased by regulation and for the automatic indexation of the maximum amount in line with inflation.)

(3) In assessing the amount of compensation in respect of the disadvantage resulting from relocation, all relevant circumstances are to be taken into account, including:

  (a) the interest in the land of the person entitled to compensation; and
  (b) the length of time the person has resided on the land (and in particular whether the person is residing on the land temporarily or indefinitely); and
  (c) the inconvenience likely to be suffered by the person because of his or her removal from the land; and
  (d) the period after the acquisition of the land during which the person has been (or will be) allowed to remain in possession of the land.

(4) Compensation is payable in respect of the disadvantage resulting from relocation if the whole of the land is acquired or if any part of the land on which the residence is situated is acquired.
(5) Only one payment of compensation in respect of the disadvantage resulting from relocation is payable for land in separate occupation.
(6) However, if more than one family resides on the same land, a separate payment may be made in respect of each family if:

  (a) the family resides in a separate dwelling house; or
  (b) the Minister responsible for the authority of the State approves of the payment.

(7) If separate payments of compensation are made, the maximum amount under subsection (2) applies to each payment, and not to the total payments.

5.3.2.2 Compensation for solatium

Consistent with s60(1), solatium (or the disadvantage resulting from relocation) only applies to the relocation of the person's principal place of residence as a result of the acquisition. As the Homestead is not subject to acquisition, Mickey is not eligible for compensation for solatium nor is Pluto as the shop is not his principal place of residence.

In the case study, House A is Minnie's principal place of residence and House B is Daisy's principal place of residence. Therefore, Minnie and Daisy are potentially eligible for compensation for solatium. Section 60(6)(a) allows for more than one payment if each family resides in a separate dwelling house, as Minnie and Daisy do.

In assessing compensation for solatium, the valuer should take into account all relevant circumstances (s60(3)) as summarised in Table 5.1.

A check of the Department's website shows that the maximum amount of compensation that an acquiring authority can provide to an owner for disadvantage is $95,010 for an acquisition on or after 1 July 2024. Valuer General NSW (2022) notes that this amount is applicable to the acquisition of a freehold residential property with leasehold interests considered on their individual merits having regard to a lessee's limited expectations of an ongoing tenure due to the short term nature of a lease agreement, less or no emotional attachment to the property or the area in which it is located and less difficulty in finding alternative accommodation and relocating.

Valuer General NSW (2022) provides a guide to amounts payable for leasehold compensation following acquisition in New South Wales which is provided in Table 5.2. It should, however, be noted that the Valuer General will assess each circumstance on its own merits.

Accordingly, based on Valuer General NSW (2022) and in the absence of other relevant factors, Minnie may be eligible for compensation for solatium of $47,505 (being 50% of $95,010) and Daisy may be eligible for compensation for solatium of $19,002 (being 20% of $95,010).

Table 5.1 Eligibility for solatium compensation

Section 60(3)	*Minnie – House A*	*Daisy – House B*
(a) the interest in the land of the person entitled to compensation	5 year lease expiring 31 December 2025	Month to month tenancy
(b) the length of time the person has resided on the land (and in particular whether the person is residing on the land temporarily or indefinitely)	Over a decade	3 years
(c) the inconvenience likely to be suffered by the person because of his or her removal from the land	Very traumatic event causing extreme anxiety	Welcomes the opportunity to relocate
(d) the period after the acquisition of the land during which the person has been (or will be) allowed to remain in possession of the land	Possession required on date of acquisition	Possession required on date of acquisition

Source: Author

Table 5.2 Leasehold eligibility for solatium compensation

Tenancy	*Percentage of the amount determined by the Minister and published in the Government Gazette*
Recently commenced occupancy (<1 year)	10%
Occupancy 1 to 5 years	20%
Occupancy 5 to 10 years	30%
Occupancy longer than 10 years	50%

Source: Valuer General NSW (2022)

5.4 Compensation for disturbance – principles

Disturbance comprises a payment of compensation for the costs incurred by the dispossessed as a result of being disturbed by the acquisition of their land by the acquiring authority. Prior to the introduction of the facilitating statutes, disturbance was a component of value in use or value to the owner. However, following the introduction of the facilitating statutes, this is no longer the case.

Traditionally, compensation for disturbance provided reimbursement for costs incurred by the dispossessed in the acquisition process such as legal costs, valuation fees, legal costs on the purchase of a replacement property, removalists, pest inspection, identification survey, building inspection, electricity and telephone reconnection, mail redirection, stamp duty and mortgage discharge and reinstatement costs for residential acquisitions, being paid precisely as such leading to compensation expressed in dollars and cents without rounding. It should be noted that the Court held in *Eureka Operations Pty Ltd v Transport for New South Wales* [2021] NSWLEC 41 that legal costs may include the costs of that which may be required (e.g. planning reports, contamination reports, etc.) so that the lawyer may advise the client.

While, over time, disturbance was extended to costs and losses (such as the loss of local goodwill following the acquisition of a business) and then, through facilitating statutes, to some valuation items as well, valuers should note that the sum of the cost items should still not be rounded but paid precisely in dollars and cents without rounding.

Jacobs (2015) notes that claims for sentimental damage, discomfort, tension or mental suffering are not permitted though job and lifestyle factors, rather than just financial factors, may be taken into account.

Newton and Conolly (2017) note the principle that disturbance is only paid as compensation where a property is acquired at a value reflecting its current use rather than its highest and best use (*Commonwealth v Milledge* (1953) 90 CLR 157).

Jacobs (2015) identifies the following provisions in facilitating statutes:

Commonwealth: *Lands Acquisition Act* 1989

Section 55(2)(c):

> In assessing the amount of compensation to which the person is entitled, regard shall be had to all relevant matters, including:
>
> . . . any loss, injury or damage, suffered or expense reasonably incurred, by the person that was, having regard to all relevant considerations, including any circumstance peculiar to the person, suffered or incurred by the person as a direct, natural and reasonable consequence of:
>
> (i) the acquisition of the interest; or
> (ii) the making or giving of the pre-acquisition declaration or certificate under section 24 in relation to the acquisition of the interest;
>
> other than any such loss, injury, damage or expense in respect of which compensation is payable under Pt VIII.

New South Wales: *Land Acquisition (Just Terms Compensation) Act* 1991

Section 55(d):

> In determining the amount of compensation to which a person in entitled, regard must be had to the following matters only (as assessed in accordance with this Division):
>
> (d) any loss attributable to disturbance;

Section 59:

> "Loss attributable to disturbance" of land means any of the following:
>
> (a) legal costs reasonably incurred by the persons entitled to compensation in connection with the compulsory acquisition of the land;
> (b) valuation fees of a qualified valuer reasonably incurred by those persons in connection with the compulsory acquisition of the land (but not fees calculated by reference to the value, as assessed by the valuer, of the land);
> (c) financial costs reasonably incurred in connection with the relocation of those persons (including legal costs but not including stamp duty or mortgage costs);
> (d) stamp duty costs reasonably incurred (or that might reasonably be incurred) by those persons in connection with the purchase of land for relocation (but not exceeding the amount that would be incurred for the purchase of land of equivalent value to the land compulsorily acquired);
> (e) financial costs reasonably incurred (or that might reasonably be incurred) by those persons in connection with the discharge of a mortgage and the execution of a new mortgage resulting from the relocation (but not exceeding the amount that would be

incurred if the new mortgage secured the repayment of the balance owing in respect of the discharged mortgage);

(f) any other financial costs reasonably incurred (or that might reasonably be incurred), relating to the actual use of the land, as a direct and natural consequence of the acquisition.

Victoria: *Land Acquisition and Compensation Act* 1986

Section 41(1)(d):

any loss attributable to disturbance

Section 40:

"Loss attributable to disturbance" means any pecuniary loss suffered by a claimant as the natural, direct and reasonable consequence of –

(a) the service upon the claimant of a notice of intention to acquire, where the Authority has refused or failed to give consent to the carrying out of improvements to the land in respect of which the notice has been served or the effecting or obtaining of any sales, transactions, licences or approvals in respect of that land; and

(b) the fact that an interest of the claimant in that land has been divested or diminished, being a pecuniary loss for which provision is not otherwise made in this Part.

Section 41(1)(d):

(1) Except as otherwise provided in this Part, in assessing the amount of compensation payable to a claimant in respect of an interest in land which is acquired under this Act, regard must be had to the following factors –

(d) any loss attributable to disturbance;

Queensland: *Acquisition of Land Act* 1967

Section 20(1), (2) and (2A) state:

(1) In assessing the compensation to be paid, regard shall in every case be had not only to the value of the land taken but also –

(b) to the claimant's costs attributable to disturbance.

(2) Compensation shall be assessed according to the value of the estate or interest of the claimant in the land taken on the date when it was taken.

(2A) However, in assessing the compensation, a contract, licence, agreement or other arrangement (a relevant instrument) entered into in relation to the land after the notice of intention to resume was served on the claimant must not be taken into consideration if the relevant instrument was entered into for the sole or dominant purpose of enabling the claimant or another person to obtain compensation for an interest in the land created under the instrument.

(5) In this section –
costs attributable to disturbance, in relation to the taking of land, means all or any of the following –

(a) legal costs and valuation or other professional fees reasonably incurred by the claimant in relation to the preparation and filing of the claimant's claim for compensation;
(b) the following costs relating to the purchase of land by a claimant to replace the land taken –
    (i) stamp duty reasonably incurred or that might reasonably be incurred by the claimant, but not more than the amount of stamp duty that would be incurred for the purchase of land of equivalent value to the land taken;
    (ii) financial costs reasonably incurred or that might reasonably be incurred by the claimant in relation to the discharge of a mortgage and the execution of a new mortgage, but not more than the amount that would be incurred if the new mortgage secured the repayment of the balance owing in relation to the discharged mortgage;
    (iii) legal costs reasonably incurred by the claimant;
    (iv) other financial costs, other than any taxation liability, reasonably incurred by the claimant;
(c) removal and storage costs reasonably incurred by the claimant in relocating from the land taken;
(d) costs reasonably incurred by the claimant to connect to any services or utilities on relocating from the land taken;
(e) other financial costs that are reasonably incurred or that might reasonably be incurred by the claimant, relating to the use of the land taken, as a direct and natural consequence of the taking of the land;
(f) an amount reasonably attributed to the loss of profits resulting from interruption to the claimant's business that is a direct and natural consequence of the taking of the land;
(g) other economic losses and costs reasonably incurred by the claimant that are a direct and natural consequence of the taking of the land.

Western Australia: *Land Administration Act* 1997

Section 241(6):

(6) Regard is to be had to the loss or damage, if any, sustained by the claimant by reason of –

(a) removal expenses;
(b) disruption and reinstatement of a business;
(c) the halting of building works in progress at the date when the interest is taken and the consequential termination of building contracts;
(d) architect's fees or quantity surveyor's fees actually incurred by the claimant in respect of proposed buildings or improvements which cannot be commenced or continued in consequence of the taking of the interest; or
(e) any other facts which the acquiring authority, the court, or the State Administrative Tribunal considers it just to take into account in the circumstances of the case.

South Australia: *Land Acquisition Act* 1969

Section 25(1)(b)(ii):

(1) The compensation payable under this Act in respect of the acquisition of land shall be determined according to the following principles:

(b) in assessing the amount referred to in paragraph (a) of this section consideration may be given to –

(ii) the loss occasioned by reason of severance, disturbance or injurious affection;

Tasmania: *Land Acquisition Act* 1993

Section 27(1)(f) and (2):

(1) In determining compensation under this Act, regard is to be had to the following matters:

(f) any disturbance relating to any loss or damage suffered, or cost reasonably incurred, by the claimant as a consequence of the taking of the subject land;

with ss2 providing for betterment offset.

Northern Territory: *Land Acquisition Act* 1978
Section 66 as read with Schedule 2, rr2(d) and 7 (in regard to compensation for resettlement):

Subject to this Schedule, in assessing the compensation payable to a claimant in respect of the acquired land the Tribunal may take into account –
. . . any loss sustained, or cost incurred, by the claimant as a natural and reasonable consequence of:

(i) the acquisition of the land; or
(ii) the service on the claimant of the notice of proposal,
for which provision is not otherwise made under this Act, other than costs incurred as a result of attending, participating in or being represented at consultations for the purposes of section 37(1) or mediation under section 37(4).

Australian Capital Territory: *Lands Acquisition Act* 1994

Section 45(2)(c)(i) and (ii), (d) and (e)(i) and (ii):

(2) In assessing the amount of compensation to which the person is entitled, regard shall be had to all relevant matters, including -

(c) any loss, injury or damage suffered, or expense reasonably incurred, by the person that was, having regard to all relevant considerations, including any circumstances peculiar to the person, suffered or incurred by the person as a direct, natural and reasonable consequence of –

(i) the acquisition of the interest; or
(ii) the making or giving of the pre-acquisition declaration or certificate under section 21 in relation to the acquisition of the interest;
other than any such loss, injury, damage or expense in respect of which compensation is payable under part 7; and

(d) subject to section 50, if the interest is limited as to time or may be terminated by another person – the likelihood of the continuation or renewal of the interest and the likely terms and conditions on which any continuation or renewal would be granted; and

(e) any legal or other professional costs reasonably incurred by the person in relation to the acquisition, including the costs of –

(i) obtaining advice in relation to the acquisition, the entitlement of the person to compensation or the amount of compensation; and
(ii) executing, producing or surrendering such documents, and making out and providing such abstracts and attested copies, as the chief executive or a person authorised under the Government Solicitor Act 1989, section 5(4) requires.

Valuers should note the precise wording of the provisions of the facilitating statute relevant to the State or Territory of dispossession, as Courts are tending towards a strict literal interpretation of statutory provisions (see, for example, the decision of the Court of Appeal in *Roads and Maritime Services v United Petroleum Pty Ltd* [2019] NSWCA 41). Where there are multiple elements to the provision, each element must be satisfied if compensation is to be payable. For example, in s59(f) of the *Land Acquisition (Just Terms Compensation) Act* 1991 (NSW)) each of the following elements must be satisfied:

- financial costs;
- reasonably incurred (or that might reasonably be incurred);
- relating to the actual use of the land; and
- as a direct and natural consequence of the acquisition.

The meaning of direct, natural and reasonable was considered by the Court in *Halwood Corporation Ltd v Roads Corporation* (1995) 89 LGERA 280 with direct meaning immediate or without intervening agency, natural meaning arising according to the usual course of things and reasonable meaning not going beyond the limit assigned by reason or not extravagant or excessive.

Newton and Conolly (2017) note that, therefore, there must be a very close connection between the event and the cost suffered in order for it to be eligible for compensation as disturbance – essentially, the cost or loss must not be too remote (*Harvey v Crawley Development Corporation* [1957] 1 QB 485 at 494).

However, losses incurred by an owner as a result of their own decisions are not recoverable, such as a decision not to develop a site which was anticipated to be but had not yet been compulsorily acquired (*Sydney Metro v G & J Drivas Pty Ltd* [2024] NSWCA 5),

Related to the requirement for a claim for disturbance compensation to be reasonable is the requirement for the dispossessed to mitigate it's claim for disturbance compensation. As stated in *Director of Buildings and Lands v Shun Fung Iron Works Ltd* [1995] 2 AC 111 at 126:

> Likewise, if a reasonable person in the position of the claimant would not have incurred, or would not incur, the expenditure being claimed, fairness does not require that the authority should be responsible for such expenditure.

5.4.1 Compensation for disturbance – businesses

On acquisition, the holder of an interest to which a business is attached may choose to:

- relocate and continue trading – in which case disturbance compensation for relocation may be payable as considered briefly in Section 4.2.2 of Chapter 4 and further, below; or
- cease trading and close down – in which case compensation for business extinguishment may be payable, being the value of the business as a going concern, as considered briefly in Section 4.2.3 of Chapter 4 and which is considered further in Section 6.4 of Chapter 6.

Such compensation for business extinguishment or relocation may payable under the disturbance provisions of the various facilitating statutes. However, some cases in New South Wales have suggested that compensation for the extinguishment of a business can only be paid as market value of the acquired interest.

While compensation for business extinguishment will be likely to be the value of the business as a going concern (see Chapter 6, Section 6.4), compensation for relocation will be dependent on the precise wording of the relevant facilitating statute with complex case law applying to the differences and similarities between relocation and reinstatement (see Jacobs, 2015, chapter 25; Newton and Conolly, 2017, pages 164–167). Given regularly changing case law, it is recommended that the valuer seek legal advice through the client when considering compensation for a complex business relocation.

Concerning relocation, the general principles of a direct, natural and reasonable consequence of the acquisition, being reasonably foreseeable, not being too remote and the requirement to mitigate also apply to compensation claims for businesses. For example, in *West v Roads and Traffic Authority (NSW)* (1995) 88 LGERA 266 the former owner claimed disturbance for the personal time spent in locating replacement premises, but the Court held that he was only entitled to disturbance compensation for the expense of obtaining accounting advice to quantify the claim for compensation (Newton and Conolly, 2017). Similarly, legal and management time spent on prosecuting a legal claim for compensation is not recoverable as disturbance but should be considered within the claim for costs arising from the legal action.

Valuer General (2024) notes the following typical relocation costs to be claimable for compensation in New South Wales:

- legal costs in relation to relocation such as preparation of a new lease;
- other professional costs relating to relocation;
- buyers/lessee's agents costs if:
  - the claimant is resident overseas and only visits Australia infrequently; or
  - the technical specifications of the property or business are of such complexity that the claimant is unable to undertake an alternative property search himself/herself; or
  - such reason as the Valuer General may consider reasonable;
- costs associated with the relocation of a lessee's improvements;
- costs of reprinting stationery;
- notification of new address to customers;
- costs of relocating telephones and internet services;
- costs of removing and installing computers;
- general removal expenses;
- Australia Post mail redirection;
- loss of business profits due to relocation. (pages 24–25)

Care is, however, required to avoid double counting between disturbance and other heads of compensation. For example, where a fitout for a specialist medical surgery that is acquired is compensated within the head of market value, a disturbance claim for the fitout of a new specialist medical surgery cannot be grounded as this would involve an element of double counting (*Matcam Pty Ltd v Kogarah Municipal Council* (1999) 105 LGERA 266) (Hyam, 2014).

Compensation for business extinguishment and business relocation may require consideration on a relative basis. In *Re Mills and Minister for Finance and Administration* (2002) 70

ALD 780, the Court held that if the costs of relocation exceeded the costs of extinguishment, compensation should be based on the cost of extinguishment. However, in *Director of Buildings and Lands v Shun Fung Ironworks Ltd* [1995] 1 All ER 846, the Court noted that *this will not always be so, and a rigid limitation . . . could lead to injustice* (853).

Further, in situations where the dispossessed business owner genuinely intends to relocate the business but at the time of claiming compensation has not yet found a suitable alternative site, an estimate of stamp duty and legal fees for the proposed purchase may be compensable as disturbance (*Macarbell Pty Ltd v Roads and Traffic Authority (NSW)* (2006) 149 LGERA 217; *Nasser v Roads and Traffic Authority (NSW)* (2006) 149 LGERA 289), the test of causation being that the cost is reasonably foreseeable (Newton and Conolly, 2017).

Considering the NSW *Land Acquisition (Just Terms Compensation) Act* 1991 as an example, disturbance compensation for business relocation is addressed by s59(1)(c) of the NSW *Land Acquisition (Just Terms Compensation) Act* 1991, being:

(c) financial costs reasonably incurred in connection with the relocation of those persons (including legal costs but not including stamp duty or mortgage costs)

which has given rise to extensive litigation, considered in Hyam (2014, pages 415–419, 444–445, Jacobs (2015, chapter 25) and Newton and Conolly (2017, pages 164–167). Further, in *Sydney Metro v C&P Automotive Engineers Pty Ltd* [2004] NSWCA 186, the Court significantly narrowed what can be claimed as disturbance for relocation with new capital building works not comprising relocation and so not generally claimable.

Further, it should be noted that in New South Wales, following recent Court decisions, claims for compensation for loss of profits by a lessee with a long term lease under s59(1)(f) of the NSW *Land Acquisition (Just Terms Compensation) Act* 1991 are uncertain and the valuer should seek legal advice on the specific circumstances of the acquisition under consideration.

5.5 Compensation for disturbance – practice

Having regard to the case study in Appendix 1, the assessment of compensation for disturbance may be considered as follows.

5.5.1 Facilitating statute

As the subject property is in New South Wales, the relevant facilitating statute is the *Land Acquisition (Just Terms Compensation) Act* 1991 which prescribes compensation for disturbance at s59:

Section 59:

"Loss attributable to disturbance" of land means any of the following:

(a) legal costs reasonably incurred by the persons entitled to compensation in connection with the compulsory acquisition of the land;
(b) valuation fees of a qualified valuer reasonably incurred by those persons in connection with the compulsory acquisition of the land (but not fees calculated by reference to the value, as assessed by the valuer, of the land);

(c) financial costs reasonably incurred in connection with the relocation of those persons (including legal costs but not including stamp duty or mortgage costs);
(d) stamp duty costs reasonably incurred (or that might reasonably be incurred) by those persons in connection with the purchase of land for relocation (but not exceeding the amount that would be incurred for the purchase of land of equivalent value to the land compulsorily acquired);
(e) financial costs reasonably incurred (or that might reasonably be incurred) by those persons in connection with the discharge of a mortgage and the execution of a new mortgage resulting from the relocation (but not exceeding the amount that would be incurred if the new mortgage secured the repayment of the balance owing in respect of the discharged mortgage);
(f) any other financial costs reasonably incurred (or that might reasonably be incurred), relating to the actual use of the land, as a direct and natural consequence of the acquisition.

In New South Wales, the way in which the Valuer General will interpret aspects of the facilitating statute is set out in *Compensation following compulsory acquisition: Guidance note* (Valuer General, 2024) with Section 3.4 addressing disturbance.

5.5.2 Compensation for disturbance

The identified interests in land may be eligible for disturbance compensation if they have been disturbed, with each being considered separately below.

5.5.2.1 Mickey

Mickey has submitted four invoices for costs associated with the compulsory acquisition, being $10,000 for legal advice on the acquisition, $5,000 for legal advice on likely success in an action in the Land and Environment Court, $5,000 for valuer's fees and $13,838 for stamp duty on two new houses.

You confirm that the costs comprising the disturbance claim relate to the houses and blacksmith's shop rather than to the paddock and meadow and so are compensable having regard to s61 of the *Land Acquisition (Just Terms) Compensation Act* 1991.

The eligibility of Mickey's claims for disturbance compensation is considered in Table 5.3.

Table 5.3 Mickey – compensation for disturbance

Invoice	*Allowed*	*Disallowed*
$10,000 for legal advice on the acquisition	$10,000 – s59(1)(a)	
$5,000 for legal advice on likely success in an action in the Land and Environment Court		$5,000 – not reasonably incurred in connection with the acquisition
$5,000 for valuer's fees	$5,000 – s59(1)(b)	
$13,838 for stamp duty on two new houses		$13,838 – no evidence of purchase of land for relocation – s59(1)(d)
Total	$15,000	$18,838

Source: Author

Mickey may also seek compensation for the partial extinguishment of his horse breeding business. However, while Mickey may have been eligible for compensation for the partial extinguishment of the horse breeding business under s59(f) (though this is doubtful in light of recent NSW case law), this claim will be precluded by *Commonwealth v Milledge* (1953) 90 CLR 157 in principle and s61 of the *Land Acquisition (Just Terms) Compensation Act* 1991 in New South Wales as Mickey is receiving market value compensation reflecting the highest and best use of the land not the current use of the land.

5.5.2.2 Minnie

Minnie has submitted three invoices for costs associated with vacating house A, being $100 for a solicitor to review the lease of her new rental house, $500 for removalists and $1,000 for her friend who helped Minnie find a new house to move into.

The eligibility of Minnie's claims for disturbance compensation is considered in Table 5.4.

5.5.2.3 Daisy

Daisy has submitted three invoices for costs associated with vacating house B, being $100 for a solicitor to review the lease of her new rental house, $500 for removalists and $2,500 for the cost of repainting the interior of her new rental house.

The eligibility of Daisy's claims for disturbance compensation is considered in Table 5.5.

Table 5.4 Minnie – compensation for disturbance

Invoice	*Allowed*	*Disallowed*
$100 for a solicitor to review the lease of her new rental house	$100 – s59(a)	
$500 for removalists	$500 – s59(c)	
$1,000 for her friend who helped Minnie find a new house to move into		$1,000 – not exceptional circumstance indicated in Section 3.4.1 Valuer General (2024) nor financial cost in the absence of an invoice
Total	$600	$1,000

Source: Author

Table 5.5 Daisy – compensation for disturbance

Invoice	*Allowed*	*Disallowed*
$100 for a solicitor to review the lease of her new rental house	$100 – s59(a)	
$500 for removalists	$500 – s59(c)	
$2,500 for the cost of repainting the interior of her new rental house		$2,500 – not reasonably incurred
Total	$600	$2,500

Source: Author

Table 5.6 Pluto – compensation for disturbance

Invoice	*Allowed*	*Disallowed*
$500 for the hire of a van to move stock	$500 – s59(f)	
$150 for a retirement sign	$150 – s59(f)	
$1,000 for losses suffered in a closing down sale		$1,000 – inconsistent with claim for extinguishment of business valued as a going concern and not a financial cost within the meaning of s59(1)(c)
$500 for setting up his business landline, fax machine and internet connection in his house		$500 – inconsistent with claim for extinguishment of business valued as a going concern
Total	$650	$1,500

Source: Author

Table 5.7 Goofy – compensation for disturbance

Invoice	*Allowed*	*Disallowed*
$50,000 for costs associated with moving his horse from the meadow to another meadow, which is within walking distance		$50,000 – reasonably incurred but manifestly exorbitant
Total	$0	$50,000

Source: Author

5.5.2.4 Pluto

Pluto has submitted four invoices for costs associated with vacating the blacksmith's shop, being $500 for the hire of a van to move stock from the shop to his home garage, $150 for an advice of retirement sign, $1,000 for losses suffered in a closing down sale and $500 for setting up his business landline, fax machine and internet connection in his home.

The eligibility of Pluto's claims for disturbance compensation is considered in Table 5.6.

As Pluto has decided, it is time to retire and will not relocate his shop elsewhere following the compulsory acquisition, Pluto may also seek compensation for the extinguishment of his blacksmith's business which shows future maintainable profits of $1,000 pa with comparable businesses trading on a capitalisation rate of 10%.

While NSW case law is evolving and compensation may be claimed as part of the market value of the lease, it is assumed that compensation is claimed as disturbance under the *Land Acquisition (Just Terms Compensation) Act* 1991. The assessment of the market value of a business is considered in Section 6.4.1 of Chapter 6 with the case study of compensation for Pluto for business extinguishment considered in Section 6.4.3.2.2.

5.5.2.5 Goofy

Goofy has submitted one invoice for the costs associated with moving his horse from the meadow to another meadow, which is within walking distance, being $50,000.

The eligibility of Goofy's claim for disturbance compensation is considered in Table 5.7.

Table 5.8 Donald – compensation for disturbance

Invoice	*Allowed*	*Disallowed*
$10,000 for loss of profits from billboard advertising for a period of 12 months after the date of acquisition		$10,000 – compensation under s59(f) precluded by Donald holding a tenancy at will and therefore having no security of tenure
Total	$0	$10,000

Source: Author

Table 5.9 Summary – disturbance compensation

Dispossessed	*Interest in land*	*Disturbance*
Mickey	Freehold	$15,000
Minnie	Leasehold – house A	$600
Daisy	Leasehold (periodic tenancy) – house B	$600
Pluto	Leasehold – shop	$650
Goofy	Leasehold (licence) – meadow	$0
Donald	Leasehold (tenancy at will) – billboard	$0
Total		**$16,850**

Source: Author

5.5.2.6 Donald

Donald has submitted one invoice for $10,000 for loss of profits from billboard advertising for a period of 12 months after the date of acquisition.

The eligibility of Donald's claim for disturbance compensation is considered in Table 5.8.

5.5.2.7 Compensation for disturbance – summary

The disturbance compensation is summarised in Table 5.9.

5.6 Summary and conclusions

Focusing on the key issues from a valuer's perspective, Chapters 1 and 2 introduce the concepts and principles of statutory valuation, with statutory valuation for acquisition considered in Chapters 3, 4, 5 and 6. Statutory valuation for rating and taxing is considered in Chapters 7 and 8, with regulatory valuation then considered in Chapters 9 and 10.

Chapter 1 introduced the principles of statutory valuation, including the concepts of market value, liberal/conservative estimates of value, highest and best use, potentiality and confirmation of a foresight. This provided the essential foundation principles of statutory valuation that are then applied in the following chapters on statutory valuation.

Chapter 2 considered the practice of statutory valuation including valuation methods with an emphasis on the comparative method of valuation, together with statutory valuation conventions such as the before and after method and the piecemeal method. This provided an understanding of the valuation methods and conventions that may be preferred for application in the following chapters on statutory valuation.

Statutory valuation for acquisition is considered in Chapters 3, 4, 5 and 6. Chapter 3 introduced the principles of statutory valuation for acquisition, including the statutory framework and process, with a focus on the concept of disregard of the scheme in compensation assessment and the distinction between concepts of value and worth. This provided an insight into concepts of fairness and equity in statutory valuation in the context of the disregard of compulsion.

Chapter 4 explored the market value bases of compensation in acquisition for which statutory valuation may be required, including market value of the interest and the related market value compensation concepts of severance and injurious affection. This provided a broad understanding of the heads of compensation for acquisition on a value basis that may arise in statutory valuation.

This chapter considered non-market value forms of compensation including special value, solatium and disturbance. Prior to the introduction of the various facilitating statutes, special value was a broad concept of compensation that added to market value, severance and disturbance to provide compensation for *value in use*, contributing to a sense of fairness in compensation with acquisition being considered to be on just terms. However, since the introduction of the various facilitating statutes, special value is compensation only for that specifically defined in the relevant statute and may only be rarely awarded.

Solatium, or payment for solace arising from loss of home, is heavily dependent on the exact wording of the facilitating statute. While the amount may be criticised for being miserly by those who have spent decades building a home and raising a family therein, payment of solatium generally contributes to the dispossessed having a sense of compensation for the loss of their home rather than just for the taking of their house.

In some respects, disturbance is a refund for costs incurred by the dispossessed as a result of the compulsory acquisition and so is awarded without rounding. However, some facilitating statutes have widened that which may be compensated by disturbance to include value related losses suffered, the determination of which requires valuer judgment. While it may be tempting to endeavour to apply the disturbance provisions of a facilitating statute as a catch-all for costs or losses not compensated elsewhere in the facilitating statute, such claims are restricted by the exact wording of the relevant clauses in that facilitating statute.

In the context of the case study adopted for chapters 3, 4, 5 and 6, the compensation due is summarised in Table 5.10.

Chapter 6 concludes the chapters considering statutory valuation for acquisition by examining compensation for other interests such as easements and mortgages, mineral rights and profits à prendre, strata title and substrata, businesses and the emerging areas of native title and of indigenous cultural loss. This provides an appreciation of the diversity of interests that may be subject to acquisition and so may require valuation for the assessment of compensation.

Statutory valuation for rating and taxing is considered in Chapters 7 and 8. Chapter 7 introduces the principles of land as the basis for taxation through land tax, stamp duty and rates, considering the general principles of the valuation of land for taxation using the New South Wales statutory framework as an example. This provides a general understanding of the often arcane and artificial basis of statutory valuation upon which Government collects very significant amounts of taxation revenue.

Chapter 8 explores the practice of statutory valuation for rating and taxing through exemptions, concessions and allowances that have evolved over the last century, together with contemporary issues such as mass appraisal, automated valuation models and artificial intelligence, the impact of major disasters on valuation for rating and taxing and the litigation process including the valuer as an expert witness. This provides an insight into the continued evolution of

Table 5.10 Summary – compensation

Dispossessed	*Interest in land*	*Market value*	*Severance*	*Injurious affection*	*Special value*	*Solatium*	*Disturbance*
Mickey	Freehold – house A	$240,000					$15,000
	Freehold – house B	$250,000					
	Freehold – shop	$175,000					
	Freehold – meadow	$375,000					
	Billboard structure	$50,000					
	Freehold – paddock	$2,625,000					
		$3,715,000	$25,000	$750,000			
Minnie	Leasehold – house A	$4,400				$47,505	$600
Daisy	Leasehold (periodic tenancy) – house B	$0				$19,002	$600
Pluto	Leasehold – shop	$5,500			$1,100		$650 plus business compensation
Goofy	Leasehold (licence) – meadow	$0					$0
Donald	Leasehold (tenancy at will) – billboard	$0					$0
Total		**$3,724,900**	**$25,000**	**$750,000**	**$1,100**	**$66,507**	**$16,850**

Source: Author

statutory valuation to address issues as they arise over time, highlighting the importance of a principles-based approach to statutory valuation

Regulatory valuation, considered in the context of the preparation of valuations for inclusion in financial reporting, is considered in Chapters 9 and 10. Chapter 9 explores the principles of regulatory valuation with regard to the inter-related roles of international accounting standards, international financial reporting standards and international valuation standards, focusing on the primacy of fair value in international financial reporting standards. This provides an understanding of the fundamental conceptual differences and bases between statutory valuation and regulatory valuation.

Finally, Chapter 10 concludes regulatory valuation by exploring the practice of valuation approaches and methods relevant for regulatory valuation such as the income approach and the cost approach and the various valuation methods there within. This provides an appreciation of the wider range of valuation approaches and methods available to the valuer for regulatory valuation reflecting that, unlike statutory valuation, there is no requirement for defence in Court (negligence claims excepted).

References

Baum A, Sams G, Ellis J, Hampson C and Stevens D (2014) *Statutory valuations*, Routledge, Abingdon

Brown, D (2009) *Land acquisition*, Lexis Nexis Butterworths, Chatswood

Hyam, A (2014) *The law affecting valuation of land in Australia*, The Federation Press, Annandale

Jacobs, M (2015) *Law of compulsory land acquisition*, Thomson Reuters, Pyrmont

Newton, G and Conolly, C (2017) *Land acquisition*, Lexis Nexis Butterworths, Chatswood

Parker, D (Ed) (2022) *Principles and practice of property valuation in Australia*, Routledge, Abingdon

Parker, D (2024) *Introduction to property valuation in Australia*, Routledge, Abingdon

Valuer General (2024) *Compensation following compulsory acquisition: guidance note*, NSW Valuer General

Valuer General of New South Wales (2022) *Determination of compensation for disadvantage resulting from residential relocation*, Valuer General of New South, Sydney

Appendix 1 – Case study

A1.1 Overview

Mickey runs a large horse breeding and racing business, based in Happytown, a small country town in NSW, comprising a 10,400 sqm block of land located on the edge of Happytown together with several hundred hectares of grazing land surrounding Happytown.

Prior to the compulsory acquisition, the 10,400 sqm block of land owned by Mickey was surrounded by Mouse Street, Duck Street and Hound Street, as shown in the before graphic (2.0). The block comprised a paddock to which horses would be brought from Mickey's surrounding grazing lands for grazing, shoeing, veterinary attention and general care, together with a homestead and a water bore, a meadow used for grazing horses with a billboard, a blacksmith's shop and two houses, each of which is separately fenced off, with the entire 10,400 sqm being zoned for medium density residential development.

The proposed acquisition will facilitate the construction, operation and maintenance of the Happytown Freeway through Mickey's block of land, as shown in the after graphic (3.0). The land to be acquired is marked by an arrow in the graphic below. The acquisition will be gazetted on 31 December 2024.

As a result of the compulsory acquisition, the portions of the block retained by Mickey north and south of the proposed Happytown Freeway will no longer meet Council's land size criteria for medium density residential development but will meet Council's criteria for low density residential development.

A1.2 Before graphic

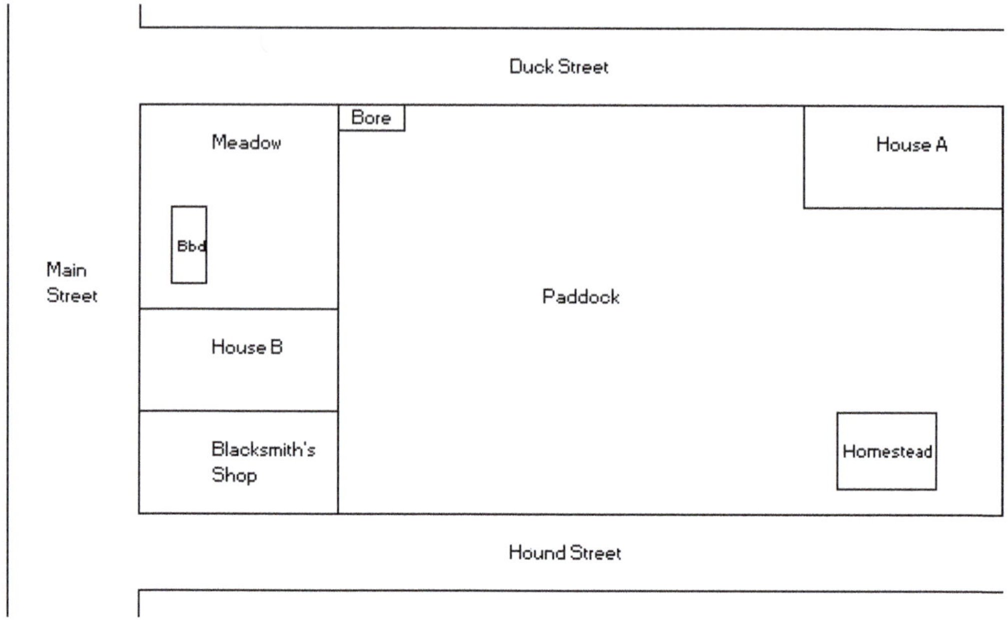

A1.3 After graphic

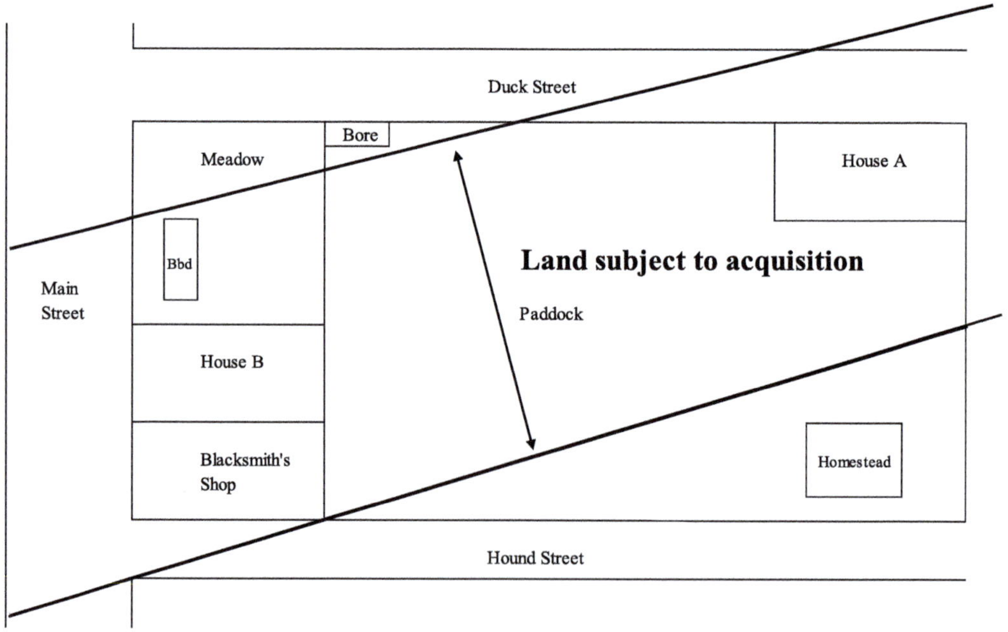

A1.4 Land areas

The respective land areas are as follows:

Area of Mickey's block	10,400 sqm
Area of homestead and paddock	7,400 sqm
Area of shop site	600 sqm
Area of house A and B sites	600 sqm each
Area of meadow	1,200 sqm
Area to be acquired	7,800 sqm
Comprising:	
Paddock	5,250 sqm
House A	600 sqm
House B	600 sqm
Shop	600 sqm
Meadow	750 sqm
Portion of the block retained north of the proposed freeway	600 sqm (450 sqm meadow, 150 sqm paddock)
Portion of the block retained south of the proposed freeway	2,000 sqm

A1.5 Parties

Mickey occupies the homestead and the paddock.

Mickey leases house A to Minnie on a five-year lease expiring 31 December 2025 at $100 pcm. Minnie has been the occupant of house A as her principal place of residence for over a decade. Minnie is very elderly and moving house is a very traumatic event causing extreme anxiety. The acquiring authority requires possession on 31 December 2024.

Mickey leases house B to Daisy on a month-to-month tenancy at $100 pcm. Daisy has been the occupant of house B as her principal place of residence for three years. Daisy was thinking of moving to a smaller residence and welcomes the opportunity to relocate. The acquiring authority requires possession on 31 December 2024.

Mickey leases the shop to Pluto on a ten-year lease expiring 31 December 2025 at $250pcm. Pluto runs a blacksmith's shop, with existing use rights, which services Mickey's horses in the paddock as well as minor metalwork jobs for residents of Happytown, showing future maintainable profits of $1,000 pa and holding stock worth $5,000. Pluto has decided it is time to retire and will not relocate his blacksmith's shop elsewhere following the compulsory acquisition.

Mickey has given a licence to Goofy to use the meadow for six months expiring 31 December 2024 at $1 pcm.

Mickey owns the billboard and has given a tenancy at will to Donald to use the billboard at $60pcm.

A1.6 Facilitating statute

The proposed acquisition notice stated that the land was to be acquired *for the purposes of the Roads Act . . . in connection with the construction, operation and maintenance of the Happytown Freeway.*

A1.7 Valuation information

Your accumulation, analysis, adjustment and application of comparable sale transactions for houses A and B indicate that the market value for each house and curtilage is $250,000.

Your accumulation, analysis, adjustment and application of comparable rental transactions for houses A and B indicate that the market rental for each house is $500 pcm.

Your accumulation, analysis, adjustment and application of comparable sale transactions for short term residential leasehold interests indicate that such interests trade on a capitalisation rate of 10%.

Your accumulation, analysis, adjustment and application of comparable rental transactions for the blacksmith's shop indicate that the market rental for the shop is $750 pcm.

Your accumulation, analysis, adjustment and application of comparable sales for freehold shops and curtilage indicate that shops trade on a capitalisation rate of 5%.

Your accumulation, analysis, adjustment and application of comparable sale transactions for short term leasehold shop interests indicate that such interests trade on a capitalisation rate of 10%.

Your analysis of the turnover of the blacksmith's shop shows future maintainable profits of $1,000 pa.

Your accumulation, analysis, adjustment and application of comparable business sales show that blacksmith's businesses trade on a capitalisation rate of 10%.

Your accumulation, analysis, adjustment and application of comparable rentals for meadow grazing indicate a market rental for 1200 sqm to be $120 pcm.

Your accumulation, analysis, adjustment and application of comparable rentals for billboards indicate a market rental to be $120 pcm.

Your accumulation, analysis, adjustment and application of comparable sales for billboard businesses indicate that billboard businesses trade on a capitalisation rate of 10%.

Your analysis of the turnover of the horse breeding and training business conducted on the land shows future maintainable profits of $10,000 pa.

Your accumulation, analysis, adjustment and application of comparable business sales show that horse breeding and training businesses trade on a capitalisation rate of 15%.

Your accumulation, analysis, adjustment and application of comparable sales for pastoral land with a water supply suitable for grazing indicate a market rate of $100psm.

Your accumulation, analysis, adjustment and application of comparable sales for low density residential development land indicate a market rate of $250psm.

Your accumulation, analysis, adjustment and application of comparable sales for medium density residential development land indicate a market rate of $500psm.

You have received advice that the replacement cost of the billboard is $250,000 and that the existing billboard has a potential remaining usable life of 10 years, having been erected 40 years ago.

A1.8 Disturbance claims

Mickey has submitted four invoices for costs associated with the compulsory acquisition, being $10,000 for legal advice on the acquisition, $5,000 for legal advice on likely success in an action in the Land and Environment Court, $5,000 for valuer's fees and $13,838 for stamp duty on two new houses.

Minnie has submitted three invoices for costs associated with vacating house A, being $100 for a solicitor to review the lease of her new rental house, $500 for removalists and $1,000 for her friend who helped Minnie find a new rental house to move into.

Daisy has submitted three invoices for costs associated with vacating house B, being $100 for a solicitor to review the lease of her new rental house, $500 for removalists and $2,500 for the cost of repainting the interior of her new rental house.

Pluto has submitted four invoices for costs associated with vacating the blacksmith's shop, being $500 for the hire of a van to move stock from the shop to his home garage, $150 for an advice of retirement sign, $1,000 for losses suffered in a closing down sale and $500 for setting up his business landline, fax machine and internet connection in his home.

Donald has submitted one invoice for $10,000 for loss of profits from billboard advertising for a period of 12 months after the date of acquisition.

Goofy has submitted one invoice for the costs associated with moving his horse from the meadow to another meadow, which is within walking distance, being $50,000.

6 Compulsory acquisition

Compensation for other interests

6.1 Introduction

This book is an introduction to two very different valuation worlds in Australia – the complex and often arcane national realm of statutory valuation and the dynamic global empire of regulatory valuation. Statutory valuation is different from regulatory valuation, with the former being based on relevant applicable statutes and being capable of challenge on appeal in Court while the latter is based on relevant applicable regulations and incapable of challenge in Court (negligence claims excepted).

As an introductory textbook on statutory and regulatory valuation, this book is a companion to the Australian valuation textbooks, *Introduction to Property Valuation in Australia* (Parker, 2024) and *Principles and Practice of Property Valuation in Australia* (Parker, 2022), which provide a wider analysis of key principles underlying property valuation and current techniques and issues in the practice of property valuation for major sectors of the Australian property market.

Statutory valuation in Australia is most commonly encountered in valuations for rating, taxing and acquisition but may arise wherever a statute prescribes a basis of or approach to valuation. Regulatory valuation is most commonly encountered in valuations for financial statements but may arise wherever a regulation prescribes a basis of or approach to valuation.

Rating and taxing of land raise significant amounts of revenue for State Governments and Councils in Australia, based on statutory valuation. For example, the NSW 2025–26 Budget Statement estimates land tax revenue for 2024–2025 to be $8.2 billion with the City of Brisbane Annual Plan and Budget 2024–2025 estimating proposed rates and utility charges to be $1.51 billion.

Acquisition of land, property and businesses commonly arises for major infrastructure projects and may be by agreement or by compulsion, based on statutory valuation. For example, such major infrastructure projects in Australia currently include the multi-billion-dollar Sydney Metro, Melbourne Suburban Rail Loop and Melbourne to Brisbane Inland Rail projects.

Valuation of land and property holdings based on regulatory valuation arises for all reporting entities in Australia that publish financial statements upon which it is reasonable to expect that users will rely for decision making. For example, at June 2025 there are over 2,000 reporting entities listed on the Australian Stock Exchange which each publish financial statements, with the financial statements for Goodman Group, as at 30 June 2024, showing the fair value of total property assets to be $20.2 billion, based on regulatory valuation.

Focusing on the key issues from a valuer's perspective, the concepts and principles of statutory valuation are introduced in Chapters 1 and 2, with statutory valuation for acquisition considered in Chapters 3, 4, 5 and 6 and statutory valuation for rating and taxing considered in Chapters 7 and 8. Regulatory valuation is then considered in Chapters 9 and 10.

DOI: 10.1201/9781032616957-6

Chapter 1 introduced the principles of value in statutory valuation, including the concepts of market value, liberal/conservative estimates of value, highest and best use, potentiality and confirmation of a foresight. This provided the essential foundation principles of statutory valuation that are then applied in the following chapters on statutory valuation.

Chapter 2 considered the practice of statutory valuation including valuation methods for statutory valuation with an emphasis on the comparative method of valuation, together with statutory valuation conventions such as the before and after method and the piecemeal method. This provided an understanding of the valuation methods and conventions that may be preferred in the following chapters on statutory valuation.

Statutory valuation for land acquisition is considered in Chapters 3, 4, 5 and 6, supported by an evolving case study in Appendix 1 of each chapter, including the principal bases of compensation being:

- compensation for market value for the interest acquired;
- compensation for severance;
- compensation for injurious affection;
- compensation for special value;
- compensation for solatium or disadvantage resulting from relocation; and
- compensation for disturbance.

Chapter 3 introduced the principles of statutory valuation for land acquisition, including the statutory framework and process, with a focus on the concept of the disregard of the scheme in compensation assessment and the distinction between concepts of value and worth. This provided an insight into concepts of fairness and equity in statutory valuation in the context of the requirement to disregard the compulsory nature of the acquisition.

Chapter 4 explored the market value bases of compensation in acquisition for which statutory valuation may be required, including market value of the interest and the related market value compensation concepts of severance and injurious affection. This provided a broad understanding of the heads of compensation for acquisition on a value basis that may arise in statutory valuation.

Chapter 5 considered non-market value forms of compensation including special value which was noted to rarely arise, payment for solace arising from loss of the primary place of residence and the concept of disturbance for costs and losses suffered as a result of the acquisition. This provided a broad understanding of the heads of compensation for acquisition that are not, necessarily, on a value basis but which may arise in statutory valuation.

This chapter concludes the chapters considering statutory valuation for acquisition by examining compensation for other interests such as easements and mortgages, mineral rights and profits à prendre, strata title and substrata, businesses and the emerging statutory valuation areas of native title and indigenous cultural loss. This provides an appreciation of the diversity of interests that may be subject to acquisition and so may require valuation for the assessment of compensation.

Statutory valuation for rating and taxing is considered in Chapters 7 and 8. Chapter 7 introduces the principles of land as the basis for taxation through land tax, stamp duty and rates, considering the general principles of the valuation of land for taxation using the New South Wales statutory framework as an example. This provides a general understanding of the often arcane and artificial basis of statutory valuation upon which Government collects very significant amounts of taxation revenue.

Chapter 8 explores the practice of statutory valuation for rating and taxing through exemptions, concessions and allowances that have evolved over the last century, together with contemporary issues such as mass appraisal, automated valuation models and artificial intelligence, the impact of major disasters on valuation for rating and taxing and the litigation process including the valuer as an expert witness. This provides an insight into the continued evolution of statutory valuation to address issues as they arise over time, highlighting the importance of a principles-based approach to statutory valuation.

Regulatory valuation, considered in the context of the preparation of valuations for inclusion in financial reporting, is considered in Chapters 9 and 10. Chapter 9 explores the principles of regulatory valuation with regard to the inter-related roles of international accounting standards, international financial reporting standards and international valuation standards, focusing on the primacy of fair value in international financial reporting standards. This provides an understanding of the fundamental conceptual differences and bases between statutory valuation and regulatory valuation.

Finally, Chapter 10 concludes regulatory valuation by exploring the practice of valuation approaches and methods relevant for regulatory valuation such as the income approach and the cost approach and the various valuation methods there within. This provides an appreciation of the wider range of valuation approaches and methods available to the valuer for regulatory valuation reflecting that, unlike statutory valuation, there is no requirement for defence in Court (negligence claims excepted).

Accordingly, this chapter will now consider:

- compensation for easements, mortgages, mineral rights and profits à prendre;
- compensation for strata title and substrata;
- compensation for businesses and
- compensation for native title and indigenous cultural loss.

6.2 Compensation for easements, mortgages, mineral rights and profits à prendre

This section seeks to guide the valuer through the valuation aspects of the key issues concerning compensation for easements, mortgages, mineral rights and profits à prendre from a valuer's viewpoint. Greater detail on the legal aspects of the key issues may be found in Jacobs (2015), Hyam (2014), Brown (2009) and Newton and Conolly (2017).

6.2.1 Compensation for easements

Easements are a proprietary right to carry out activities on land owned by another subject to the terms of the easement. Private easements benefit land (the dominant tenement) while burdening other land (the servient tenement) or they can be public easements in favour of a public authority (easements in gross).

Acquisition of an easement commonly occurs where a utility is proposing to install water, drainage, gas or sewerage pipes, or underground or overhead power transmission lines. In most cases, the acquiring authority is not seeking to acquire the land itself but seeks to impose on the title a right to pass over or under the land in a specified manner and to restrict the use of the land by the owner within the boundaries of the easement.

Further, an acquisition of a freehold interest in land can also extinguish existing easements. This can affect both the acquired land and the land that has the benefit of the easement.

Related to easement are rights of light, mineral rights and profits à prendre, where an adjoining owner is prevented from restricting the flow of light to a property or where a party has the right to take something from or off another's land such as sand or gravel.

6.2.1.1 Compensation for easements – principles

Newton and Conolly (2017) identify three ways in which the acquisition of an easement may arise:

- the acquiring authority may resume land in which there is already an existing easement vested in the neighbouring land. This gives the neighbour an entitlement to compensation;
- the land resumed may have the right to an easement over neighbouring land. The acquisition is unlikely to affect the easement because it is not being extinguished and
- the acquiring authority may wish to acquire an easement (or more accurately impose the easement on the title) without acquiring other interests such as the freehold or a leasehold interest in the land.

In addition, certain facilitating statutes have specific provisions in relation to easements. These are:

- New South Wales – *Land Acquisition (Just Terms Compensation) Act* 1991 – s62(1) provides that if the land acquired consists only of an easement or right to use land under the surface for the construction or maintenance of works (such as a tunnel, pipe or conduit), compensation is not payable except for actual damage done;
- Queensland – *Acquisition of Land Act* 1967 – s6 authorises a constructing authority to take an easement; and
- Western Australia – *Land Administration Act* 1997 – s195 authorises an acquiring authority to take an easement with s196 authorising the creation of public access easements, such as a right of way for the use and benefit of the public at large.

6.2.1.2 Compensation for easements – practice

Where the acquiring authority is acquiring land over which there is an existing easement, Jacobs (2015) notes that the assessment of compensation for the acquisition of the easement should be considered from the subjective perspective of the easement holder/beneficiary. For example, a farmer who holds an easement permitting access over adjoining land which is acquired may seek compensation for the cost of creating a new form of access or, if that is not possible, for the diminution in the value of his land due to loss of access.

Where the acquiring authority is seeking to create a new easement, such as for the installation of drainage and sewerage pipes, underground or overhead power transmission lines, pipelines, support or for rights of way (such as to access to highways or railway lines), the principal issue is the degree to which the land over which the easement is acquired will be reduced in value because of the easement.

For example, an easement for a drainage pipe may run along the boundary of a suburban house property but will likely require that nothing be built on top of the easement reservation, lest access is required for maintenance of the drainage pipe, so constraining the homeowner's use of the land within the easement reservation.

Similarly, an overhead transmission line easement may run across a large field used for growing crops and to avoid risk of fire or interfering with the power line the easement will most likely require that nothing is built around the transmission towers and may require the area beneath the transmission lines to be kept clear, so constraining the farmer's use of the land within the easement reservation.

Where an easement is imposed by the acquiring authority on the title of the land, generally, the effect will be to reduce the value of that land. The compensation may be for a reduction in the market value of the land within the easement and potentially include severance and/or injurious affection. It may also give rise to specific costs being incurred by the landowner which may be claimed as disturbance. As the Court noted in *Joyce v Northern Electrical Authority of Queensland* (1974) 1 QLCR 171:

> For practical purposes it becomes a matter of assessing the extent to which he has been disadvantaged as the natural and reasonable consequence of the taking of the easement.

In *Besmaw Pty Ltd v Sydney Water Corporation* [2001] NSWLEC 15, the Court considered a claim for compensation following the compulsory acquisition of an easement for a sewer main over the applicant's land. While the land was used for the extraction of sand, the owner also had development consent for redevelopment as a tourist resort which required access to an adjoining road. The easement terms required the owner to obtain consent for alterations to the surface level of the land subject to the easement, with the Court holding that the terms of the easement did not guarantee that the applicant could construct the necessary access without the Corporation's consent which it was entitled to decline. The Court, therefore, awarded $1.5 million compensation for disturbance under s59(f) of the *Land Acquisition (Just Terms Compensation) Act* 1991 (Hyam, 2014).

Further, the compulsory acquisition of several new easements, the interaction between each and their interaction with existing easements from the viewpoint of compensation was considered by the Court in *Penrith City Council v Sydney Water Corporation* [2009] NSWLEC 2.

Where available, the use of comparable sales and the before and after method or the piecemeal method are preferable to determine compensation. Jacobs (2015) notes that s41(3) of the Victorian *Land Acquisition and Compensation Act* 1986 requires the application of the before and after method.

However, other than in Victoria, in the absence of comparable sales, the diminution in value is usually expressed as a percentage of market value with the quantum of the percentage varying depending on the level of invasiveness of the easement being acquired. Having regard to case law on percentage deductions, the NSW Valuer General issued a guide for valuers which is reproduced in Table 6.1 (Valuer General, 2024a).

It should, however, be noted that, in some circumstances, the blight attributable to the impact of the creation of the easement for which the acquisition is required may be modest, nominal or practically non-existent other than as *a blot on the title to land.* In such circumstances, the acquiring authority and the land holder may negotiate and agree a modest (say $10,000), nominal (say $1,000) or token (say $1) payment.

On occasion, other forms of easement may be acquired requiring the assessment of compensation. In *Bligh Consulting Pty Ltd v Ausgrid* [2016] NSWLEC 75, the Court considered an easement for rock anchors and temporary easements for crane swing and scaffolding. While no compensation was payable for rock anchors due to s62 of the *Land Acquisition (Just Terms Compensation) Act* 1991, comparable transactions for scaffolding easements were scarce leading the Court to award compensation of $50,000, being $25,000 market value and $25,000 injurious affection, for the minimal impact of the scaffolding easement. Based on the evidence

Table 6.1 Percentage of market value for impact of easements

Range	*Commonly*	*Impact of easement*
0–10%	0%	Negligible impact – for example: - where easement has no physical impact on land - where the sole impact of the easement is as a blot on title - where the easement is fully sub-surface (e.g. a tunnel, pipe or rock anchor), noting the provisions in s62 of the Act
10–30%	10%	Limited impact – for example: - where the easement will traverse open space that can in any case only be used by the landowner as open space (e.g. a public park) - where above-ground or overhead structures result in loss of visual amenity - where the easement otherwise does not significantly impinge upon the landowner's use of land
30–60%	50%	Low to moderate – for example: - visible surface impact (e.g. such as pipes above the surface of the ground or electricity substations) that diminish visual amenity for the landowner and create inconvenience but still allow the land to be used
60–80%	80%	Moderate to high impact – for example: - other restrictions on use or rights under the easement that significantly impinge the landowner's use of land
80–100%	100%	Severe to total impact – for example: - where the landowner is legally unable to build in the easement area (e.g. beneath an electricity transmission line) - where the terms of the easement authorise the land to be occupied by a structure such as a pumping station or electricity pylon which completely forestalls the landowner's use of the that land

Source: Valuer General (2024a)

of one comparable transaction provided by one of the expert valuers, compensation for market value of $400,000 was awarded for the crane swing easement.

6.2.2 *Compensation for mortgages*

Land that is acquired may be subject to a mortgage, where the mortgagee (lender) has lent money to the mortgagor (borrower) secured by the land. The effect of the acquisition is to terminate the mortgage. Accordingly, in the event of acquisition, the mortgagee has an interest in land and is entitled to claim compensation for the value of the mortgage.

6.2.2.1 Compensation for mortgages – principles

A mortgage registered on title is an interest in land and, in theory, if it is extinguished it gives rise to potential compensation to the holder.

A mortgage is a form of security giving rights to the mortgagee (lender) to exercise a power of sale if the mortgagor (borrower) defaults on a loan provided by the mortgagee. Mortgages are typically a form of security that is allowed to be imposed on the title of the land under the loan agreement.

Newton and Conolly (2017) note that all facilitating statutes include provisions for the payment of compensation for the mortgagee's interest in the land:

Commonwealth	
Lands Acquisition Act 1989	ss53, 62, 64–69, 84, 92
Victoria	
Land Acquisition and Compensation Act 1986	ss68–73
New South Wales	
Land Acquisition (Just Terms Compensation) Act 1991	s65
Queensland	
Land Acquisition Act 1967	s32
Western Australia	
Land Administration Act 1997	s251
South Australia	
Land Acquisition Act 1969	s26
Tasmania	
Land Acquisition Act 1993	s34
Northern Territory	
Lands Acquisition Act 1978	Sch 2 r 10
Australian Capital Territory	
Lands Acquisition Act 1994	ss39, 43, 51–58, 77

However, if compensation is paid to a mortgagee this will reduces the amount of compensation payable to the registered proprietor. Generally, mortgagees do not make claims for compensation arising from an acquisition of land that extinguishes the mortgage. The reason being that they rely on the loan agreement to require the repayment of the loan. If the land is acquired, typically the loan agreement will require the immediate payment of any outstanding amount owed to the mortgagee.

Given the practice of including the mortgagee's interest within the assessment of compensation for market value, the valuer is only likely to need to refer to the exact wording of the facilitating statute in unusual circumstances.

6.2.2.2 Compensation for mortgages – practice

Generally, the assessment of market value of the acquired land will be on the basis that there is no mortgage registered on title. While an acquisition notice is required to be served on both the landowner and mortgagee, it is common for mortgagees to not make a claim for compensation and to just rely on the loan agreement to enforce payment of any outstanding amount.

Sometimes a mortgagee may make a claim and require direct payment of compensation to the mortgagee based on the amount outstanding under the mortgage. In such a situation, this can be paid out by the acquiring authority with the balance of compensation for market value being paid to the landowner.

6.2.3 Compensation for mineral rights

Jacobs (2015) notes that it may be necessary to determine compensation for mineral rights in the following cases:

- where mining takes place on the property acquired; or
- where mining does not take place on the property acquired but there is the potential for mining; or
- where a dominant tenement enjoys mineral rights over an easement tenement and the dominant tenement is acquired.

The valuation of an active mine is a specialist form of valuation that should only be undertaken by valuers with relevant expertise and experience. Where land is acquired with the potential for mining, care should be taken to value the acquired land as land with potentiality for mining rather than valuing the land and mineral rights separately.

6.2.4 Compensation for profits à prendre

Though rare in Australia, à profit a prendre is a legal right to enter another person's land and take something from that land such as trees or soil. Jacobs (2015) notes that an equitable profit à prendre has been held to constitute an interest in land. The value of such an interest, if acquired, would be dependent on the profitability of that removed.

6.3 Compensation for strata and substrata

This section seeks to guide the valuer through the valuation aspects of the key issues concerning compensation for strata title and substrata from a valuer's viewpoint. Greater detail on the legal aspects of the key issues may be found in Jacobs (2015), Hyam (2014), Brown (2009) and Newton and Conolly (2017).

Strata title interests in property may be subject to acquisition including air right interests (such as advertising billboards), conventional strata title property interests (apartments, office suites, retail shops and so on) and substratum interests (such as tunnels and underground car parks).

6.3.1 Compensation for strata title

Interests in land comprising strata that may be eligible for compensation in the event of acquisition include both above-ground interests (such as a strata title apartment) and below-ground interests (such as an underground car park).

6.3.1.1 Compensation for above-ground strata title – principles

Above-ground strata title interests include conventional strata title apartments, offices, shops, car parks and so on, as well as community title property and air right interests such as billboards and antennae. The same principles of valuation apply to above-ground strata interests as apply to non-strata interests.

It is common in major infrastructure projects for an entire strata title building to be acquired comprising a large number of individual lots (such as apartments and/or offices and/or shops and/or car parking spaces). Each of the lots will be individually acquired by the acquiring authority from each lot owner with a question then arising about the acquisition of the common property. While, as a general principle, each acquired lot may be considered to include within the acquisition that lot's beneficial entitlement to use of the common property, the valuer should seek legal advice concerning the specific law in the State or Territory of acquisition.

6.3.1.2 Compensation for above-ground strata title – practice

As interests in land, valuation practice for the valuation of above-ground strata interests is the same as that which applies to non-strata interests. For conventional strata title apartments, offices, shops, car parks and so on, the comparative method of valuation is likely to be most relevant with income methods of valuation being potentially relevant for office and shop strata title interests and for air right interests such as billboards and antennae.

6.3.2 Compensation for substrata

Sub-surface or below ground interests, being substratum interests, can arise in a variety of forms such as a railway tunnel passing beneath a building. As, generally, the fee simple interest in land extends from the centre of the earth to the heavens (*usque ad coelum et ad inferos*), that part of the land which is required for the construction of a railway tunnel will require acquisition from the freeholder. Other substrata interests may include strata title interests (such as underground car parks) and easements (for drainage, water and sewerage, rock anchors and so on).

6.3.2.1 Compensation for below ground strata – principles

While most facilitating statutes are silent concerning below ground interests, Newton and Conolly (2017) note that the *Land Acquisition (Just Terms Compensation) Act* 1991 (NSW) includes s62 in respect of tunnels and s6 of the *Acquisition of Land Act* 1967 (Qld) and s195 of the *Land Administration Act* 1997 (WA) authorise the taking of easements.

6.3.2.2 Compensation for below ground strata – practice

Each type of acquisition and the relevant facilitating statute will need to be examined in detail to determine the principles for assessing compensation. Where a sub-surface easement is acquired (or imposed on title) below ground, the facts need to be examined. An easement 50 m below the surface is unlikely to give rise to any compensation. But if it is near the surface and affects the development potential of the land or the support of the surface of the land, then compensation may be payable. The compensation may be for market value and severance and/or injurious affection, depending on the effect of the easement on the land.

Generally, in the absence of comparable sales, the diminution in value is usually expressed as a percentage with the quantum of the percentage varying depending on the level of impact of the easement being acquired.

Certain jurisdictions have specific provisions dealing with sub-surface acquisitions. For example, the acquisition of a below ground interest in New South Wales for a tunnel is addressed in s62(2) of the *Land Acquisition (Just Terms Compensation) Act* 1991:

Section 62:

(1) If the land compulsorily acquired under this Act consists only of an easement, or right to use land, under the surface for the construction and maintenance or works (such as a tunnel, pipe or conduit for the conveyance of water, sewage or electrical cables), compensation is not payable except for actual damage done in the construction of the work or caused by the work.

(2) If the land under the surface is compulsorily acquired under this Act for the purpose of constructing a tunnel, compensation is not payable (subject to subsection (1)) unless:

  (a) the surface of the overlying soil is disturbed, or
  (b) the support of that surface is destroyed or injuriously affected by the construction of the tunnel, or
  (c) any mines or underground workings in or adjacent to the land are thereby rendered unworkable or are injuriously affected.

(3) If the land compulsorily acquired under this Act consists of or includes an easement or right to use the surface of the land for the construction and maintenance of works (such as canals,

> drainage, stormwater channels, electrical cables, openings or ventilators), the easement or right is (unless the acquisition notice otherwise provides) taken to include a power, from time to time, to enter the land for the purpose of inspection and for carrying out of any additions, renewals or repairs. Compensation under this Part is payable accordingly.

Effectively, no compensation will be payable for the acquisition of a below ground interest for a tunnel unless the overlying soil is disturbed, support is damaged or mines are impacted. Given the use of large boring machines to cut tunnels far beneath the land, disturbance to soil or damage to support is unlikely to be evident until a significant period of time after the acquisition, with the full extent not known until after the tunnel has been cut and the land has settled.

Therefore, in order to complete determinations of compensation for tunnel acquisitions in a timely manner, the following approach is adopted by the NSW Valuer General:

> The Valuer General generally issues a determination of nil compensation immediately after gazettal.
>
> (Valuer General, 2024b, section 1.6.2)

Further, in both *Pennant Hills Golf Club Ltd v Roads and Traffic Authority (NSW)* (1997) 96 LGERA 164 and *Bligh Consulting Pty Ltd v Ausgrid* [2016] NSWLEC 75, the Courts held that, in New South Wales, easements for rock anchors situated below the ground and not causing disturbance to the surface of the land above were within s62 and not compensable.

6.4 Compensation for businesses

This section seeks to guide the valuer through the valuation aspects of the key issues concerning compensation for businesses attached to an interest in land from a valuer's viewpoint. Greater detail on the legal aspects of the key issues may be found in Jacobs (2015), Hyam (2014), Brown (2009) and Newton and Conolly (2017).

The first question to determine is whether the business has a separate compensable interest in the land. Some businesses (sole traders for example) are operated by the owner of the land and have no separate interest in land. Other businesses may be incorporated and occupy the land under a lease (either registered or unregistered). If the business has an interest in land, then it has a separate entitlement to compensation to that of the owner. For example, a leasehold shop and business may be acquired with the compensation for the business forming part of the compensation for the leasehold interest.

On acquisition, the owner of the interest to which the business is attached may choose to:

- relocate and continue trading – in which case compensation for relocation may be payable as considered in Section 4.2.2 of Chapter 4 and Section 5.4.1 of Chapter 5; or
- cease trading and close down – in which case compensation for business extinguishment may be payable, being the value of the business as a going concern, as considered in Section 4.2.3 of Chapter 4 and in Section 6.4.1.

6.4.1 Compensation for business extinguishment – principles

Newton and Conolly (2017) note that the facilitating statutes do not directly refer to situations where the acquisition causes, or partly contributes to, the extinguishment of a business with claimants needing to rely for compensation on provisions governing disturbance, loss or damage. Generally, on extinguishment, compensation would reflect the value of the business as a going concern.

For example, in *Bligh v Minister Administering Environmental Planning and Assessment Act* [2011] NSWLEC 220, the Court ruled that the dispossessed was entitled to compensation for business extinguishment under the disturbance provision of s59(f) of the *Land Acquisition (Just Terms Compensation) Act* 1991. However, there have been a number of recent cases in NSW which raise doubts that disturbance costs are payable for the value of a business if it is forced to close as a result of acquisition. These cases suggest that compensation is payable for the value of the business as part of the market value of the interest held by the business (either freehold or leasehold) (see, e.g. *Alexandria Landfill Pty Ltd v Transport for NSW* [2020] NSWCA 165).

Each factual situation in each jurisdiction will need to be examined in detail and potentially legal advice obtained on how to claim compensation when a business is extinguished by an acquisition.

Generally, the value of a business reflects the potential to earn profits and create goodwill in the business. Newton and Conolly (2017) define goodwill as:

> The goodwill of a business is the advantage of the reputation and connection formed with customers together with circumstances, whether of habit or otherwise, which tend to make such connection permanent. It is the value of that element of profitability that arises from specific business connections related to the location of the premises acquired or the personality of the proprietor. It is the benefit and advantage of the good name, reputation and connection.
>
> (page 173)

Goodwill is not a stand-alone item for the purposes of compensation but forms part of the claim for the legal interest to which it is attached, such as the freehold, leasehold, licence or other interest.

Goodwill may attach to the land (local goodwill) and to the business proprietor (personal goodwill). In the event of business extinguishment, local goodwill and personal goodwill may be lost and so may be compensable.

However, consistent with the sale of a business (and its interest in land) by a business owner on retirement (being the market value of the business as a going concern), a business buyer may pay local goodwill for the benefits attached to the location (which is embedded in market value) but may not pay for personal goodwill as the proprietor is retiring and will no longer attend the business.

Therefore, potentially foregoing personal goodwill through choosing to extinguish the business on acquisition rather than preserving personal goodwill by relocating the business is a choice for the dispossessed.

As will be considered further in Section 6.4.2, the market value of a business as a going concern is usually a function of the future maintainable profit after tax multiplied by the capitalisation rate or price earnings ratio – with local goodwill being embedded within the multiplier.

6.4.2 Compensation for business extinguishment – practice

As noted in Chapter 2, Section 2.3.2.1, for a property valuer, the valuation of a business as a going concern on extinguishment may be likely to arise in the context of a small business holding a leasehold or license interest in a property, as a large business as a going concern is likely to be valued by a business valuer or an accountant. However, the valuer should ensure that they have the appropriate experience and knowledge to undertake the valuation of a small business.

It should be noted that, where a business operates under insecure tenure such as a short term lease, a holding over lease or a lease which can be terminated at short notice by the lessor, the interest in land is not likely to have any market value. Thus, the value of the business as a going concern is likely to be very low or of no value. For example, in NSW, a licence is unlikely to give rise to a compensable interest in land (see *Dial A Dump Industries Pty Ltd v Roads and Maritime Services* [2017] NSWCA 73]).

Generally, to have any value, the business must have a secure and long term interest in land, through attachment to either the owner of the land or the lessee under a long-term secure (and preferably registered) lease.

Concerning the assessment of the value of a business as a going concern, Hyam (2014) notes that the capitalisation of profit or 'the going concern basis' was accepted for the valuation of profit-earning undertakings in *Hobart Bridge Co Ltd v Tasmania* (1945) 9 The Valuer 62 at 63, *Eastaway v Commonwealth* (1951) 84 CLR 328 and *Re Electricity Commission (Balmain Electric Light Co Purchase) Act 1950* (1955) 1 LGRA 49 at 81 and, more recently, in *Emerald Quarry Industries Pty Ltd v Commissioner of Highways (SA)* (1979) 142 CLR 351.

Lonergan (2022) notes four principal valuation methodologies for consideration in the valuation of a business as a going concern:

- capitalisation of future maintainable profit;
- capitalisation of future maintainable dividends;
- discounted cash flow or
- asset based valuations.

In the context of a small business holding a leasehold or license interest in a property, the capitalisation of future maintainable profits is likely to be the preferred method of valuation:

Future maintainable profit after tax
× 100/capitalisation rate
= Value of business as a going concern

or:

Future maintainable profit after tax
× Price earnings ratio
= Value of business as a going concern

though cross checking with the use of a second method is advisable.

Therefore, this requires the valuer to ascertain the future maintainable profit after tax for the business and the appropriate capitalisation rate or price earnings ratio.

Lonergan (2022) notes that, to determine the future maintainable profit after tax for a business, the earnings of the business should be normalised by excluding abnormal, non-recurring and non-arms-length transactions but including any unrecognised costs and income that are not reflected in the results.

Items that may require adjustment include:

- compensation to owners and spouses, so that these are representative of arms-length arrangements;
- depreciation rates;
- non-recurring items, such as business interruptions, discontinued operations, abnormal market conditions;
- rental or lease payments on an arms-length basis;
- any related party transactions on an arms-length basis (eg management fees);
- any non-operating income or expenses that are not related to the principal business; and
- interest on shareholder loans other than on an arms-length basis. (Lonergan, 2022)

As the aim is to determine future maintainable profit, the following require analysis and reflection:

- trend and cyclicality of current profits;
- previous and (reasonably) forecast normalised profits;
- industry and economic trends;
- risk factors affecting the business;
- liquidity position;
- security of tenure in respect of premises, agency rights, franchises, licences;
- asset support to the capitalised value;
- debt/equity ratios;
- continuity and experience of management; and
- experiences (success/failure) of entities of a similar nature. (Lonergan, 2022)

The capitalisation rate or price earnings ratio should be sourced from comparable business transactions, following the general principles adopted in the comparative method of valuation. It is, therefore, preferable to identify and accumulate sales of businesses in similar industries, of a similar size, in a similar geographic location and with similar firm specific characteristics (such as management) and then analyse, adjust and apply to the subject business. For example, adjustment will be required for size (smaller multiples apply to smaller entities), growth prospects (higher multiples for higher growth prospects) and negotiability (lower multiples apply to less negotiable assets) (Lonergan, 2022).

6.4.3 Case study – compensation for business

Appendix 1 contains a case study which will be used to illustrate compensation for business extinguishment following compulsory acquisition.

6.4.3.1 Facilitating statute

As the subject property in Appendix 1 is in New South Wales, the relevant facilitating statute is the *Land Acquisition (Just Terms Compensation) Act* 1991 which prescribes compensation for disturbance at s59(f):

Section 59:

'Loss attributable to disturbance' of land means any of the following:

(f) any other financial costs reasonably incurred (or that might reasonably be incurred), relating to the actual use of the land, as a direct and natural consequence of the acquisition.

It should be noted that recent cases in New South Wales raise doubts that disturbance costs are payable for the value of a business if it is forced to close as a result of acquisition. These cases suggest that compensation is payable for the value of the business as part of the market value of the interest held by the business (either freehold or leasehold). However, for the purposes of this example, it is assumed that compensation may be claimed as disturbance.

6.4.3.2 Business compensation

To determine the value of a business as a going concern, the preferred method would be to use the capitalisation of future maintainable profit. This requires the valuer to determine the future maintainable profit after tax and the relevant capitalisation rate or price earnings ratio as detailed in Section 6.4.2.

The businesses that are being impacted by the acquisition include:

- Mickey's horse breeding and racing business as conducted in the paddock;
- Pluto's blacksmith's business and
- Donald's billboard business.

6.4.3.2.1 MICKEY – HORSE BREEDING BUSINESS

Mickey occupies the homestead and cares for horses in the paddock. Mickey's homestead and paddock comprised 7,400 sqm before the acquisition and 2,150 sqm after the acquisition. Therefore, the loss of 5,250 sqm of paddock may be expected to significantly impact Mickey's horse breeding and racing business.

Your analysis of the turnover of the horse breeding and training business show future maintainable profits of $10,000 pa. Your accumulation, analysis, adjustment and application of comparable business sales show that horse breeding and training businesses trade on a capitalisation rate of 15%.

However, for market value compensation, Mickey was compensated for the market value of the highest and best use of the paddock, being the market rate for medium density residential development land of $500 psm. As Mickey has been compensated for the highest and best use value of the paddock, rather than $100 psm for agricultural land with a water supply, s61 of the *Land Acquisition (Just Terms Compensation) Act* 1991 precludes a claim for further compensation (*El Boustani v Minister Administering the Environmental Planning and Assessment Act* 1979 [2012] NSWLEC 266).

6.4.3.2.2 PLUTO – BLACKSMITH'S SHOP BUSINESS

Pluto runs a blacksmith's shop which shows future maintainable profits of $1,000 pa and holds stock worth $5,000. Pluto has decided it is time to retire and will not relocate his blacksmith's shop elsewhere following the compulsory acquisition.

Pluto is entitled to compensation for extinguishment of his business as a going concern, for which the preferred valuation method would be the capitalisation of future maintainable profit:

Future maintainable profit after tax
× 100/capitalisation rate
= Value of business as a going concern

Your analysis of the turnover of the shop shows future maintainable profit of $1,000 pa. Your accumulation, analysis, adjustment and application of comparable business sales show that blacksmith's businesses trade on a capitalisation rate of 10%:

$1,000 pa future maintainable profits after tax
× 10 capitalisation rate
= $10,000
+ $5,000 stock (reflecting sale as a going concern basis of value)
= $15,000 value of business as a going concern

This provides an assessment of compensation for Pluto for disturbance (business extinguishment) of $15,000.

6.4.3.2.3 DONALD – BILLBOARD BUSINESS

Mickey has given a tenancy at will to Donald to use the billboard which provides Donald with no security of tenure. Accordingly, as Donald's billboard business has no certainty of future maintainable profits after tax, the insecure tenure means the interest in land and the business operated on that land has no value on extinguishment of the tenancy.

6.4.3.2.4 BUSINESS COMPENSATION – SUMMARY

The disturbance compensation for business extinguishment is summarised in Table 6.2.

6.5 Compensation for native title and indigenous cultural loss

This section seeks to guide the valuer through the valuation aspects of the key issues concerning compensation for native title and indigenous cultural loss from a valuer's viewpoint. Greater detail on the legal aspects of the key issues may be found in Jacobs (2015), Hyam (2014), Brown (2009) and Newton and Conolly (2017).

6.5.1 Compensation for native title

The determination of compensation for native title rights is a complex and specialist area of valuation. The following is a brief, general introductory guide only and valuers considering

Table 6.2 Summary – disturbance (business extinguishment) compensation

Dispossessed	*Interest in land*	*Disturbance (business extinguishment)*
Mickey	Freehold	$0
Minnie	Leasehold – house A	$0
Daisy	Leasehold (periodic tenancy) – house B	$0
Pluto	Leasehold – shop	$15,000
Goofy	Leasehold (licence) – meadow	$0
Donald	Leasehold (tenancy at will) – billboard	$0
Total		**$15,000**

Source: Author

undertaking such a determination should ensure that they have the necessary skills and experience or undertake the valuation in association with another valuer who has the necessary skills and experience, consistent with API *Rules of Professional Conduct* (1.6) (API, 2024) and RICS *Valuation – Global Standards (Red Book) PS 2 (Ethics, competency, objectivity and disclosures)* (RICS, 2024).

6.5.1.1 Compensation for native title – principles

Jacobs (2015) cites the following statutes that concern native title land:

Commonwealth	*Native Title Act* 1993
New South Wales	*Native Title (NSW) Act* 1994
South Australia	*Native Title (SA) Act* 1994
Tasmania	*Native Title (Tas) Act* 1994
Western Australia	*Native Title (State Provisions) Act* 1999
Victoria	*Native Titles Validation Act* 1994
Queensland	*Native Title (Qld) Act* 1993
Northern Territory	*Valuation (Native Title) Act* 1994
Australian Capital Territory	*Native Title Act* 1994

and the following facilitating statutes concerning the acquisition of native title land:

Commonwealth	*Native Title Act* 1993 – s24MD(2),
New South Wales	*Land Acquisition (Just Terms Compensation) Act* 1991 – s7A, s12(5), s37A, s54(2)
South Australia	*Land Acquisition Act* 1969 – s25(1)(a) and (b), (2) and (3)
Tasmania	*Land Acquisition Act* 1993 – s5A
Western Australia	*Land Administration Act* 1997 – s152–158
Northern Territory	*Land Acquisition Act* 1978 – s5A, s31A, s31C, s50A, s59(2)(a)-(d), s81A

Generally, the Commonwealth legislation and the States and Territories legislation are linked so that the dispossessed cannot be worse off under the latter than the former following acquisition.

Further, some facilitating statutes such as s37A(2) of the *Land Acquisition (Just Terms Compensation) Act* 1991 (NSW) and s50A of the *Land Acquisition Act* 1978 (NT) provide for compensation in a form other than money which may include community buildings and services.

The level of compensation payable for acquired non-exclusive native title rights was considered in *Griffiths v Northern Territory (No 3)* (2016) 337 ALR 362 where the Court characterised native title as a communal bundle of rights as opposed to an individual proprietary right. The Court determined that the freehold value of the land set the upper limit of compensation, but where the rights are non-exclusive some reduction from the freehold value is necessary. Further, the Court noted that even though the disposal of native title rights on the open market is not possible, this did not necessarily mean that a significant discount should be applied.

6.5.1.2 Compensation for native title – practice

The practice of valuing native title rights is continuously evolving and may vary by State and Territory such that it is essential that the valuer has the necessary skills and experience to undertake such a valuation. While the level of discount for the non-exclusive nature of native title rights may vary case by case and by State and Territory, it is common in New South Wales for native title land acquired to be compensated at 50% of the rate applicable to comparable freehold land.

6.5.2 Compensation for indigenous cultural loss

In addition to compensation for the loss of land, dispossessed indigenous groups may also be eligible for compensation for cultural loss and/or loss of spiritual attachment to the land acquired (non-economic loss) following the acquisition of the land.

As to what comprises cultural loss and/or loss of spiritual attachment, Edelman J observed in *Northern Territory v Griffiths* (2019) HCA 7 that:

> 312 Expressed more fully, it is compensation for the value of the loss of attachment to country and rights to live on, and to gain spiritual and material sustenance from, the land.

The determination of compensation for indigenous cultural loss and/or loss of spiritual attachment is currently one of the cutting edge issues in statutory valuation globally as those countries with indigenous populations grapple with determining what is fair and just compensation.

Remarkably little research has been done into the valuation of cultural loss and/or loss of spiritual attachment in Australia, with the principal reference being *Review Of Forms Of Cultural Loss And The Process And Method For Quantifying Compensation For Compulsory Acquisition* prepared by the NSW Valuer General (Valuer General, 2022) which forms the basis for *Compensation for cultural loss arising from compulsory acquisition – guidance note* (Valuer General, 2024c). The following is based on and draws heavily from Valuer General (2022) and Valuer General (2024c) which are hereby acknowledged.

6.5.2.1 Compensation for indigenous cultural loss – principles

Division 5 of Part 2 of the Commonwealth *Native Title Act* 1993 provides for the determination of compensation for acts affecting native title. The key provision of that Division is s51, which provides:

> **Just compensation**
>
> (1) Subject to subsection (3), the entitlement to compensation under Division 2, 2A, 2B, 3 or 4 is an entitlement on just terms to compensate the native title holders for any loss, diminution, impairment or other effect of the act on their native title rights and interests.
>
> **Acquisition under compulsory acquisition law**
>
> (2) If the act is the compulsory acquisition of all or any of the native title rights and interests of the native title holders, the court, person or body making the determination of compensation on just terms may, subject to subsections (5) to (8), in doing so have regard to any principles or criteria for determining compensation set out in the law under which the compulsory acquisition takes place.

noting the reference to *rights and interests*.

For the purposes of New South Wales, s7A of the *Land Acquisition (Just Terms) Act* 1991 provides:

7A Authority empowered to acquire native title

(1) An authority of the State that is authorised by law to acquire land by compulsory process in accordance with this Act is authorised to acquire native title rights and interests in relation to the land in the same way that other interests in the land may be acquired.
(2) For the purposes of any such acquisition of native title rights and interests, the authority of the State is, despite any provision of this or any other Act to the contrary, authorised to comply with any relevant procedure under the Commonwealth Native Title Act for a valid acquisition of those rights and interests.

Accordingly, in New South Wales, compensation for cultural loss and/or spiritual loss arises as compensation for the acquisition of *native title rights and interests in relation to the land.*

Currently, the leading authority for the law pertaining to compensation for acts affecting native title is *Northern Territory v Griffiths* (2019) HCA 7 (*Timber Creek*). While it is the only case which has articulated the principles pertaining to the determination of compensation for acts affecting native title, it provided limited guidance on the process or methodology for valuing cultural loss and/or spiritual loss.

Timber Creek is a remote community in the Northern Territory, approximately 600kms south of Darwin, with the case concerning the loss of native title rights over an area of 127 hectares in and around the town of Timber Creek. The Court awarded $1.3 million compensation for cultural loss, providing supporting principles but no guidance as to the basis for determination of the quantum of compensation, listing the following rights in accordance with traditional laws and customs of the claim group (14):

(1) to travel over, move about and have access to the land;
(2) to hunt, fish and forage on the land;
(3) to gather and use the natural resources of the land such as food, medicinal plants, wild tobacco, timber, stone and resin;
(4) to have access to and use the natural water of the land;
(5) to live on the land, to camp, to erect shelters and structures;
(6) to engage in cultural activities, conduct ceremonies, to hold meetings, to teach the physical and spiritual attributes of places and areas of importance on or in the land, and to participate in cultural practices related to birth and death, including burial rights;
(7) to have access to, maintain and protect sites of significance on the application area;
(8) to share or exchange subsistence and other traditional resources obtained on or from the land (but not for any commercial purposes).

Settlement of compensation for cultural loss and/or spiritual loss is often agreed confidentially between the parties, providing very little comparable evidence as to the monetary value attributed to various aspects of cultural loss and/or spiritual loss.

In *Bandjalung Aboriginal Corporation Prescribed Body Corporate RNTBC obh of the Bandjalung People v Transport for NSW* [2020] NSWLEC 1008, the Court gave effect to a conciliation conference ordered under s34(1) of the *Land and Environment Court Act* 1979 where the parties agreed the level of compensation. While no explanation is provided as to how the amount was quantified, the initial offer of $9,080 was increased to $42,000 plus interest following conciliation. As the negotiations occurred post *Timber Creek*, the parties may have been informed by the outcome and principles of that case.

6.5.2.2 Compensation for indigenous cultural loss – practice

Valuer General (2022) identifies that the forms of cultural loss may include, but not be limited to, the following, with the examples listed being indicative rather than exhaustive:

Access

Cultural loss associated with travelling over, moving about and having access to the land.

Residence

Cultural loss associated with living on the land, to camp, erecting shelters and structures.

Activities

Cultural loss associated with gathering and using the natural resources of the land such as food, medicinal plants, wild tobacco, timber, stone and resin.
Cultural loss associated with hunting, gathering and fishing and foraging on the land.
Cultural loss associated with having access to and use of the natural water of the land.
Cultural loss associated with having access to, maintaining and protecting sites of significance on the land.
Cultural loss associated with providing, sharing or exchanging subsistence and other traditional resources obtained on or from the land with family and others (but not for any commercial purposes).

Practices

Cultural loss associated with engaging in cultural activities and social uses of the land.
Cultural loss associated with conducting ceremonies and rituals. Cultural loss associated with holding meetings.
Cultural loss associated with holding family and community cultural gatherings.
Cultural loss associated with transmitting knowledge and stories to younger generations by teaching the physical and spiritual attributes of places and areas of importance on or in the land.
Cultural loss associated with passing on resource use and practical skills.
Cultural loss associated with participating in cultural practices related to birth and death, including burial rights.

Ecology

Cultural loss associated with impact on plants, animals and the ecosystem, which may include species of totemic or other cultural importance.
Cultural loss associated with impact on water quality and access which may have broader cultural impacts.
Cultural loss associated with collateral ecological impacts on adjacent lands and waters.
Cultural loss associated with impact on cultural responsibility to care for country from an ecological perspective.

Sites

Cultural loss associated with ability to look after and speak for culturally significant sites and areas (including, but not limited to, middens, ceremonial sites, habitation sites, song lines and including impact on impacted sites beyond a site's physical footprint).
Cultural loss associated with physical access to sites and areas of significance.
Cultural loss associated with physical damage to sites and areas of significance including, but not limited to, loss of artefacts.
Cultural loss associated with collateral damage or detriment to the wider zone around sites and areas of significance in the surrounding area.
Cultural loss associated with damage to the land impacting on wider regions of cultural significance.
Cultural loss associated with impact on significant tracks or pathways with associated stories crossing the land or surrounding region.
Cultural loss associated with lack of cultural monitors and loss of ability to identify artefacts or sites which may have been at the locations of the acquisition.
Cultural loss associated with impact on spiritual landscape.

Trauma

Emotional, gut-wrenching pain and deep or primary emotions.
Distress and anxiety caused by reason of loss of part of country and sites of significance, which is manifested by deep or primary emotions of hurt, shame and worry.
Feelings of shame arising from a sense of failed responsibility under traditional laws to look after, speak for and uphold care obligations in respect of the land.
Spiritual and emotional distress arising from any damage to country including damage to significant sites, areas or ecological values.
Feelings of loss of identity.
Intergenerational loss of what otherwise would have been inalienable rights to country.

Progressive impairment.

Progressive chipping away causing detriment to the ability to exercise rights and diminishing cultural and spiritual connection to country.
Incremental and cumulative loss of feeling of connection to country.
Earlier acts which were not compensable but punched holes in what could be likened to a single large painting, being a single and coherent pattern of belief in relation to a far wider area of land.

Concerning the difficulty in quantifying compensation for cultural loss and/or spiritual loss, Valuer General (2022) quotes from a review by Associate Professor Garrick Small which stated:

> Attempting to estimate the money value of spiritual attachment is equivalent to the absurdity of attempting to measure the length of blue, or the weight of sweet.

Having regard to *Timber Creek*, Valuer General (2022) proposed the following principles for the quantification of compensation for cultural loss and/or spiritual loss:

- that the compensation amount be determined as compensation for cultural loss rather than as compensation for special value under s57 of the Act;
- that the compensation amount be determined as compensation for cultural loss rather than as compensation for solatium under s60 of the Act;

- that the compensation amount be determined as monetary value being market value having regard to notions of worth through the proposed valuation methodology;
- that the compensation amount reflect a loss in perpetuity and compensated by the payment of a single capital sum for all generations;
- that the compensation amount be determined on an in globo basis without division by form of cultural loss or by parcel of land acquired (unless only one parcel acquired), with the apportionment or distribution of the award to be resolved among those who had suffered the loss;
- that the compensation amount has regard to the extent to which related areas have been impacted;
- that the compensation amount disregards the change in size of the claimant group over time;
- that the compensation amount disregards the size of the acquired land;
- that the compensation amount disregards the compensation amount for economic loss;
- that the compensation amount have regard to other determinations for cultural loss made by the Valuer General, the Courts or by Valuer Generals of other States and Territories;
- that the compensation amount be determined intuitively; and
- that the compensation amount be an amount that would be considered appropriate, fair and just in the Australian community.

In order to determine compensation, Valuer General (2022) suggested the following valuation methodology:

- that the forms and number of forms of cultural loss be identified (acknowledging that they are potentially innumerable) through the claimant's submitted claim, through conference with the claimant and through submitted evidence;
- that the significance of each of the forms of cultural loss be identified through conference with the claimant, preferably following invitation to country by the claimant;
- that the significance of each of the forms of cultural loss be expressed by the Valuer General as none (not applicable), low (less significant), medium (significant) or high (very significant);
- that the Valuer General have regard to the whole of the evidence, including the claimant's statement of claim and supporting evidence, the acquiring authority's list of issues and any consultant advice that the Valuer General may seek, within the context of the following conceptual diagram when intuitively determining compensation for cultural loss:

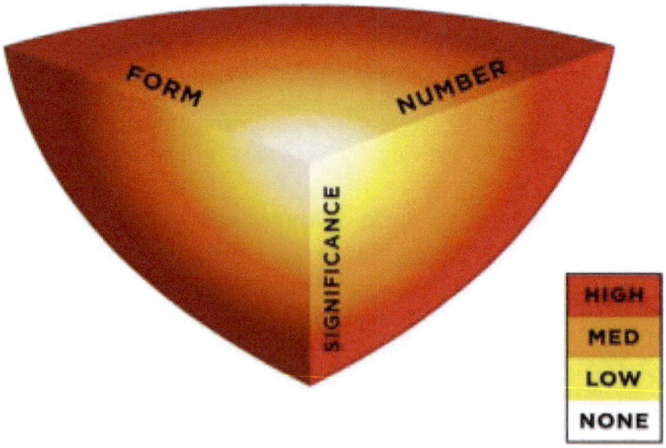

such that a wide range of forms of cultural loss which are many in number and very significant would support the highest level of compensation with fewer forms, lower number and lesser significance supporting lower levels of compensation.

A fundamental tension arises from the provisions of the *Land Acquisition (Just Terms Compensation) Act* 1991, which focuses on notions of value, and the inherent nature of cultural loss and/or spiritual loss, which focuses on notions of worth. For the foreseeable future, the amounts agreed between acquiring authorities and indigenous groups as compensation for cultural loss and/or spiritual loss may be likely to be assessments of worth to the indigenous groups subject to the constraint from *Timber Creek* of being amounts that would be considered appropriate, fair and just in the Australian community.

Notions of value are based on transactions and, in the fullness of time, a body of information concerning amounts agreed as compensation for different forms of cultural loss and/or spiritual loss will emerge to provide the valuer with transactional evidence with which to determine compensation for cultural loss and/or spiritual loss.

6.6 Summary and conclusions

Focusing on the key issues from a valuer's perspective, Chapters 1 and 2 introduced the concepts and principles of statutory valuation, with statutory valuation for acquisition considered in Chapters 3, 4, 5 and 6. Statutory valuation for rating and taxing is considered in Chapters 7 and 8, with regulatory valuation then considered in Chapters 9 and 10.

Chapter 1 introduced the principles of statutory valuation, including the concepts of market value, liberal/conservative estimates of value, highest and best use, potentiality and confirmation of a foresight. This provided the essential foundation principles of statutory valuation that are then applied in the following chapters on statutory valuation.

Chapter 2 considered the practice of statutory valuation including valuation methods with an emphasis on the comparative method of valuation, together with statutory valuation conventions such as the before and after method and the piecemeal method. This provided an understanding of the valuation methods and conventions that may be preferred for application in the following chapters on statutory valuation.

Statutory valuation for acquisition is considered in Chapters 3, 4, 5 and 6. Chapter 3 introduced the principles of statutory valuation for acquisition, including the statutory framework and process, with a focus on the concept of disregard of the scheme in compensation assessment and the distinction between concepts of value and worth. This provided an insight into concepts of fairness and equity in statutory valuation in the context of the disregard of compulsion.

Chapter 4 explored the market value bases of compensation in acquisition for which statutory valuation may be required, including market value of the interest and the related market value compensation concepts of severance and injurious affection. This provided a broad understanding of the heads of compensation for acquisition on a value basis that may arise in statutory valuation.

Chapter 5 considered non-market value forms of compensation including special value which was noted to rarely arise, payment for solace arising from loss of home which is heavily dependent on the exact wording of the facilitating statute and the concept of disturbance for costs and losses suffered as a result of the acquisition. This provided a broad understanding of the heads of compensation for acquisition that are not, necessarily, on a value basis but that may arise in statutory valuation.

This chapter concluded the chapters considering statutory valuation for acquisition by examining compensation for other interests such as easements and mortgages, mineral rights and profits à prendre, strata title and substrata, businesses and the emerging areas of native title and of indigenous cultural loss. This provides an appreciation of the diversity of interests that may be subject to acquisition and so may require valuation for the assessment of compensation.

While illustrating the diversity of interests, such interests may not arise as commonly as interests such as freehold and leasehold but their entitlement to compensation means that they cannot be ignored in the compulsory acquisition compensation process.

The wide range of forms of easements resulted in a wide range of judicial decisions on compensation that are challenging for a practising valuer to comprehend. Therefore, the New South Wales Valuer General's guidance on percentage compensation for differing levels of impact provides a handy practical guide for the valuer. While the determination of compensation for mineral rights, profits à prendre and strata title interests largely depends on the application of general statutory valuation principles, compensation for substrata interests is somewhat more complex and varies between States and Territories.

For both the valuation of businesses as a going concern and the assessment of compensation for indigenous cultural loss, it is important that the valuer recognise the limits of their competence. The valuation of a small business holding a leasehold or licence interest in a property (such as a baker's shop or car mechanics workshop) is more likely to be within the competence of a valuer whereas the valuation of a large business as a going concern is more likely to be within the competence of an accountant. Similarly, an instruction to determine an assessment of compensation for indigenous cultural loss is unlikely to be within the expertise of most valuers and so should be either declined or only undertaken with the assistance of a valuer expert in such assessment.

In the context of the case study adopted for Chapters 3, 4, 5 and 6, the compensation due is summarised in Table 6.3.

Hence, the sum of compensation payable in the case study, under the relevant heads of compensation in the NSW *Land Acquisition (Just Terms Compensation) Act* 1991, is $4,599,357. Readers will note that there are many components of compensation ranging from large amounts to small amounts which may only be determined by a rigorous and logical application of the relevant facilitating statute and case law to the facts of the specific acquisition.

Statutory valuation for rating and taxing is considered in Chapters 7 and 8. Chapter 7 introduces the principles of land as the basis for taxation through land tax, stamp duty and rates, considering the general principles of the valuation of land for taxation using the New South Wales statutory framework as an example. This provides a general understanding of the often arcane and artificial basis of statutory valuation upon which government collects very significant amounts of taxation revenue.

Chapter 8 explores the practice of statutory valuation for rating and taxing through exemptions, concessions and allowances that have evolved over the last century, together with contemporary issues such as mass appraisal, automated valuation models and artificial intelligence, the impact of major disasters on valuation for rating and taxing and the litigation process including the valuer as an expert witness. This provides an insight into the continued evolution of statutory valuation to address issues as they arise over time, highlighting the importance of a principles-based approach to statutory valuation.

Regulatory valuation, considered in the context of the preparation of valuations for inclusion in financial reporting, is considered in Chapters 9 and 10. Chapter 9 explores the principles of regulatory valuation with regard to the inter-related roles of international accounting standards, international financial reporting standards and international valuation standards, focusing on the primacy of fair value in international financial reporting standards. This provides an understanding of the fundamental conceptual differences and bases between statutory valuation and regulatory valuation.

Finally, Chapter 10 concludes regulatory valuation by exploring the practice of valuation approaches and methods relevant for regulatory valuation such as the income approach and the

Table 6.3 Summary – compensation

Dispossessed	*Interest in land*	*Market value*	*Severance*	*Injurious affection*	*Special value*	*Solatium*	*Disturbance*
Mickey	Freehold – house A	$240,000					$15,000
	Freehold – house B	$250,000					
	Freehold – shop	$175,000					
	Freehold – meadow	$375,000					
	Billboard structure	$50,000					
	Freehold – paddock	$2,625,000					
		$3,715,000	$25,000	$750,000			
Minnie	Leasehold – house A	$4,400				$47,505	$600
Daisy	Leasehold (periodic tenancy) – house B	$0				$19,002	$600
Pluto	Leasehold – shop	$5,500			$1,100		$15,650
Goofy	Leasehold (licence) – meadow	$0					$0
Donald	Leasehold (tenancy at will) – billboard	$0					$0
Total		**$3,724,900**	**$25,000**	**$750,000**	**$1,100**	**$66,507**	**$31,850**

Source: Author

cost approach and the various valuation methods there within. This provides an appreciation of the wider range of valuation approaches and methods available to the valuer for regulatory valuation reflecting that, unlike statutory valuation, there is no requirement for defence in Court (negligence claims excepted).

References

API (2024) *Rules of professional conduct*, Australian Property Institute Limited, Deakin

Brown, D (2009) *Land acquisition*, Lexis Nexis Butterworths, Chatswood

Hyam, A (2014) *The law affecting valuation of land in Australia*, The Federation Press, Annandale

Jacobs, M (2015) *Law of compulsory land acquisition*, Thomson Reuters, Pyrmont

Lonergan, W (2022) Business and intangible asset valuation, in D Parker (Ed) *Principles and practice of property valuation in Australia*, Routledge, Abingdon

Newton, G and Conolly, C (2017) *Land acquisition*, Lexis Nexis Butterworths, Chatswood

Parker, D (Ed) (2022) *Principles and practice of property valuation in Australia*, Routledge, Abingdon

Parker, D (2024) *Introduction to property valuation in Australia*, Routledge, Abingdon

RICS (2024) *RICS valuation – global standards (red book) PS 2 (ethics, competency, objectivity and disclosures)*, RICS, London

Valuer General (2022) *Review of forms of cultural loss and the process and method for quantifying compensation for compulsory acquisition*, Valuer General, Sydney

Valuer General (2024a) *Compensation following compulsory acquisition*, Valuer General, Sydney

Valuer General (2024b) *Compensation following compulsory acquisition of a substratum interest: guidance note*, Valuer General, Sydney

Valuer General (2024c) *Compensation for cultural loss arising from compulsory acquisition – guidance note*, Valuer General, Sydney

Appendix 1 – Case study

A1.1 Overview

Mickey runs a large horse breeding and racing business, based in Happytown, a small country town in NSW, comprising a 10,400 sqm block of land located on the edge of Happytown together with several hundred hectares of grazing land surrounding Happytown.

Prior to the compulsory acquisition, the 10,400 sqm block of land owned by Mickey was surrounded by Mouse Street, Duck Street and Hound Street, as shown in the before graphic (2.0). The block comprised a paddock to which horses would be brought from Mickey's surrounding grazing lands for grazing, shoeing, veterinary attention and general care, together with a homestead and a water bore, a meadow used for grazing horses with a billboard, a blacksmith's shop and two houses, each of which is separately fenced off, with the entire 10,400 sqm being zoned for medium density residential development.

The proposed acquisition will facilitate the construction, operation and maintenance of the Happytown Freeway through Mickey's block of land, as shown in the after graphic (3.0). The land to be acquired is marked by an arrow in the graphic below. The acquisition will be gazetted on 31 December 2024.

As a result of the compulsory acquisition, the portions of the block retained by Mickey north and south of the proposed Happytown Freeway will no longer meet Council's land size criteria for medium density residential development but will meet Council's criteria for low density residential development.

A1.2 Before graphic

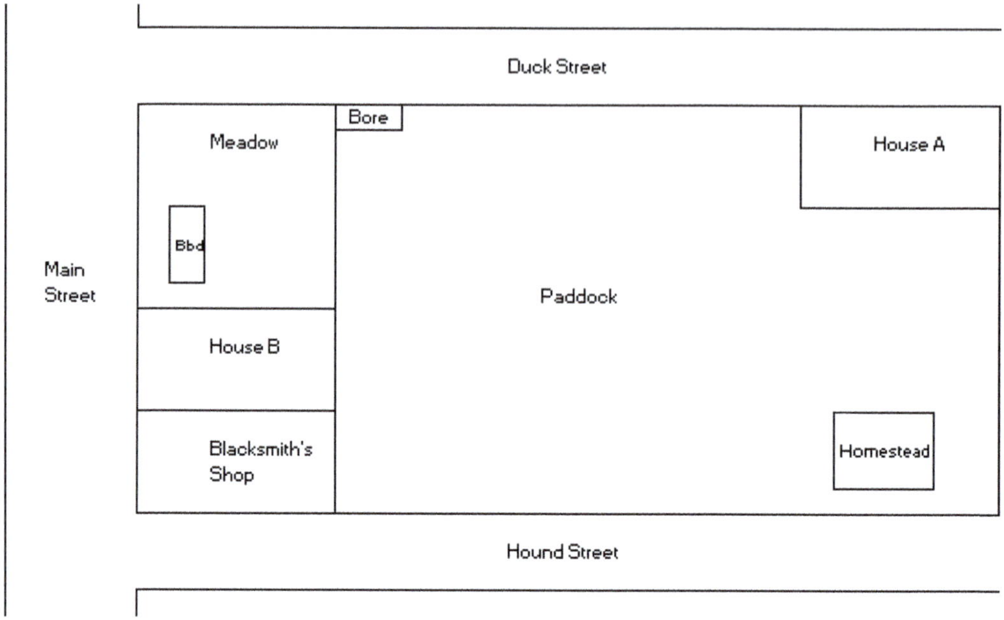

A1.3 After graphic

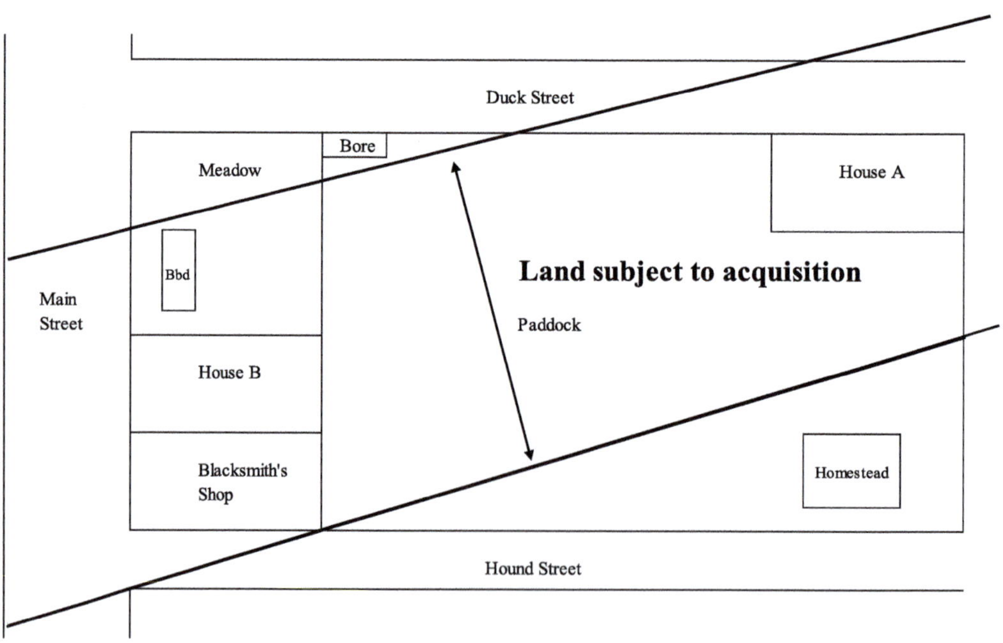

A1.4 Land areas

The respective land areas are as follows:

Area of Mickey's block	10,400 sqm
Area of homestead and paddock	7,400 sqm
Area of shop site	600 sqm
Area of house A and B sites	600 sqm each
Area of meadow	1,200 sqm
Area to be acquired	7,800 sqm
Comprising:	
Paddock	5,250 sqm
House A	600 sqm
House B	600 sqm
Shop	600 sqm
Meadow	750 sqm
Portion of the block retained north of the proposed freeway	600 sqm (450 sqm meadow, 150 sqm paddock)
Portion of the block retained south of the proposed freeway	2,000 sqm

A1.5 Parties

Mickey occupies the homestead and the paddock.

Mickey leases house A to Minnie on a five-year lease expiring 31 December 2025 at $100 pcm. Minnie has been the occupant of house A as her principal place of residence for over a decade. Minnie is very elderly and moving house is a very traumatic event causing extreme anxiety. The acquiring authority requires possession on 31 December 2024.

Mickey leases house B to Daisy on a month-to-month tenancy at $100 pcm. Daisy has been the occupant of house B as her principal place of residence for 3 years. Daisy was thinking of moving to a smaller residence and welcomes the opportunity to relocate. The acquiring authority requires possession on 31 December 2024.

Mickey leases the shop to Pluto on a ten-year lease expiring 31 December 2025 at $250pcm. Pluto runs a blacksmith's shop, with existing use rights, which services Mickey's horses in the paddock as well as minor metalwork jobs for residents of Happytown, showing future maintainable profits of $1,000 pa and holding stock worth $5,000. Pluto has decided it is time to retire and will not relocate his blacksmith's shop elsewhere following the compulsory acquisition.

Mickey has given a licence to Goofy to use the meadow for six months expiring 31 December 2024 at $1 pcm.

Mickey owns the billboard and has given a tenancy at will to Donald to use the billboard at $60 pcm.

A1.6 Facilitating statute

The proposed acquisition notice stated that the land was to be acquired *for the purposes of the Roads Act . . . in connection with the construction, operation and maintenance of the Happy-town Freeway.*

A1.7 Valuation information

Your accumulation, analysis, adjustment and application of comparable sale transactions for houses A and B indicate that the market value for each house and curtilage is $250,000.

Your accumulation, analysis, adjustment and application of comparable rental transactions for houses A and B indicate that the market rental for each house is $500 pcm.

Your accumulation, analysis, adjustment and application of comparable sale transactions for short term residential leasehold interests indicate that such interests trade on a capitalisation rate of 10%.

Your accumulation, analysis, adjustment and application of comparable rental transactions for the blacksmith's shop indicate that the market rental for the shop is $750 pcm.

Your accumulation, analysis, adjustment and application of comparable sales for freehold shops and curtilage indicate that shops trade on a capitalisation rate of 5%.

Your accumulation, analysis, adjustment and application of comparable sale transactions for short term leasehold shop interests indicate that such interests trade on a capitalisation rate of 10%.

Your analysis of the turnover of the blacksmith's shop shows future maintainable profits of $1,000 pa.

Your accumulation, analysis, adjustment and application of comparable business sales show that blacksmith's businesses trade on a capitalisation rate of 10%.

Your accumulation, analysis, adjustment and application of comparable rentals for meadow grazing indicate a market rental for 1200 sqm to be $120 pcm.

Your accumulation, analysis, adjustment and application of comparable rentals for billboards indicate a market rental to be $120 pcm.

Your accumulation, analysis, adjustment and application of comparable sales for billboard businesses indicate that billboard businesses trade on a capitalisation rate of 10%.

Your analysis of the turnover of the horse breeding and training business conducted on the land shows future maintainable profits of $10,000 pa.

Your accumulation, analysis, adjustment and application of comparable business sales show that horse breeding and training businesses trade on a capitalisation rate of 15%.

Your accumulation, analysis, adjustment and application of comparable sales for pastoral land with a water supply suitable for grazing indicate a market rate of $100 psm.

Your accumulation, analysis, adjustment and application of comparable sales for low density residential development land indicate a market rate of $250 psm.

Your accumulation, analysis, adjustment and application of comparable sales for medium density residential development land indicate a market rate of $500 psm.

You have received advice that the replacement cost of the billboard is $250,000 and that the existing billboard has a potential remaining usable life of 10 years, having been erected 40 years ago.

A1.8 Disturbance claims

Mickey has submitted four invoices for costs associated with the compulsory acquisition, being $10,000 for legal advice on the acquisition, $5,000 for legal advice on likely success in an action in the Land and Environment Court, $5,000 for valuer's fees and $13,838 for stamp duty on two new houses.

Minnie has submitted three invoices for costs associated with vacating house A, being $100 for a solicitor to review the lease of her new rental house, $500 for removalists and $1,000 for her friend who helped Minnie find a new rental house to move into.

Daisy has submitted three invoices for costs associated with vacating house B, being $100 for a solicitor to review the lease of her new rental house, $500 for removalists and $2,500 for the cost of repainting the interior of her new rental house.

Pluto has submitted four invoices for costs associated with vacating the blacksmith's shop, being $500 for the hire of a van to move stock from the shop to his home garage, $150 for an advice of retirement sign, $1,000 for losses suffered in a closing down sale and $500 for setting up his business landline, fax machine and internet connection in his home.

Donald has submitted one invoice for $10,000 for loss of profits from billboard advertising for a period of 12 months after the date of acquisition.

Goofy has submitted one invoice for the costs associated with moving his horse from the meadow to another meadow, which is within walking distance, being $50,000.

7 Rating and land tax

Principles

7.1 Introduction

This book is an introduction to two very different valuation worlds in Australia – the complex and often arcane national realm of statutory valuation and the dynamic global empire of regulatory valuation. Statutory valuation is different from regulatory valuation, with the former being based on relevant applicable statutes and being capable of challenge on appeal in Court while the latter is based on relevant applicable regulations and incapable of challenge in Court (negligence claims excepted).

As an introductory textbook on statutory and regulatory valuation, this book is a companion to the Australian valuation textbooks, *Introduction to Property Valuation in Australia* (Parker, 2024) and *Principles and Practice of Property Valuation in Australia* (Parker, 2022), which provide a wider analysis of key principles underlying property valuation and current techniques and issues in the practice of property valuation for major sectors of the Australian property market.

Statutory valuation in Australia is most commonly encountered in valuations for rating, taxing and acquisition but may arise wherever a statute prescribes a basis of or approach to valuation. Regulatory valuation is most commonly encountered in valuations for financial statements but may arise wherever a regulation prescribes a basis of or approach to valuation.

Rating and taxing of land raise significant amounts of revenue for State Governments and Councils in Australia, based on statutory valuation. For example, the NSW 2025–26 Budget Statement estimates land tax revenue for 2024–2025 to be $8.2 billion with the City of Brisbane Annual Plan and Budget 2024–2025 estimating proposed rates and utility charges to be $1.51 billion.

Acquisition of land, property and businesses commonly arises for major infrastructure projects and may be by agreement or by compulsion, based on statutory valuation. For example, such major infrastructure projects in Australia currently include the multi-billion-dollar Sydney Metro, Melbourne Suburban Rail Loop and Melbourne to Brisbane Inland Rail projects.

Valuation of land and property holdings based on regulatory valuation arises for all reporting entities in Australia that publish financial statements upon which it is reasonable to expect that users will rely for decision making. For example, at June 2025 there are over 2,000 reporting entities listed on the Australian Stock Exchange which each publish financial statements, with the financial statements for Goodman Group, as at 30 June 2024, showing the fair value of total property assets to be $20.2 billion, based on regulatory valuation.

Focusing on the key issues from a valuer's perspective, the concepts and principles of statutory valuation are introduced in Chapters 1 and 2, with statutory valuation for acquisition

DOI: 10.1201/9781032616957-7

considered in Chapters 3, 4, 5 and 6 and statutory valuation for rating and taxing considered in Chapters 7 and 8. Regulatory valuation is then considered in Chapters 9 and 10.

Chapter 1 introduced the principles of value in statutory valuation, including the concepts of market value, liberal/conservative estimates of value, highest and best use, potentiality and confirmation of a foresight. This provided the essential foundation principles of statutory valuation that are then applied in the following chapters on statutory valuation.

Chapter 2 considered the practice of statutory valuation including valuation methods for statutory valuation with an emphasis on the comparative method of valuation, together with statutory valuation conventions such as the before and after method and the piecemeal method. This provided an understanding of the valuation methods and conventions that may be preferred in the following chapters on statutory valuation.

Statutory valuation for land acquisition is considered in Chapters 3, 4, 5 and 6, supported by an evolving example in Appendix 1 of each chapter. Chapter 3 introduced the principles of statutory valuation for land acquisition, including the statutory framework and process, with a focus on the concept of the disregard of the scheme in compensation assessment and the distinction between concepts of value and worth. This provided an insight into concepts of fairness and equity in statutory valuation in the context of the requirement to disregard the compulsory nature of the acquisition.

Chapter 4 explored the market value bases of compensation in acquisition for which statutory valuation may be required, including market value of the interest and the related market value compensation concepts of severance and injurious affection. This provided a broad understanding of the heads of compensation for acquisition on a value basis that may arise in statutory valuation.

Chapter 5 considered non-market value forms of compensation including special value, which was noted to rarely arise, payment for solace arising from loss of the primary place of residence and the concept of disturbance for costs and losses suffered as a result of the acquisition. This provided a broad understanding of the heads of compensation for acquisition that are not, necessarily, on a value basis but which may arise in statutory valuation.

Chapter 6 concluded the chapters considering statutory valuation for acquisition by examining compensation for other interests such as easements and mortgages, mineral rights and profits à prendre, strata title and substrata, businesses and the emerging statutory valuation areas of native title and indigenous cultural loss. This provided an appreciation of the diversity of interests that may be subject to acquisition and so may require valuation for the assessment of compensation.

Statutory valuation for rating and taxing is considered in Chapters 7 and 8. This chapter introduces the principles of land as the basis for taxation through land tax, stamp duty and rates, considering the general principles of the valuation of land for taxation using the New South Wales statutory framework as an example. This provides a general understanding of the often arcane and artificial basis of statutory valuation upon which Government collects very significant amounts of taxation revenue.

Chapter 8 explores the practice of statutory valuation for rating and taxing through exemptions, concessions and allowances that have evolved over the last century, together with contemporary issues such as mass appraisal, automated valuation models and artificial intelligence, the impact of major disasters on valuation for rating and taxing and the litigation process including the valuer as an expert witness. This provides an insight into the continued evolution of statutory valuation to address issues as they arise over time, highlighting the importance of a principles-based approach to statutory valuation.

Regulatory valuation, considered in the context of the preparation of valuations for inclusion in financial reporting, is considered in Chapters 9 and 10. Chapter 9 explores the principles of regulatory valuation with regard to the inter-related roles of international accounting standards, international financial reporting standards and international valuation standards, focusing on the primacy of fair value in international financial reporting standards. This provides an understanding of the fundamental conceptual differences and bases between statutory valuation and regulatory valuation.

Finally, Chapter 10 concludes regulatory valuation by exploring the practice of valuation approaches and methods relevant for regulatory valuation such as the income approach and the cost approach and the various valuation methods there within. This provides an appreciation of the wider range of valuation approaches and methods available to the valuer for regulatory valuation reflecting that, unlike statutory valuation, there is no requirement for defence in Court (negligence claims excepted).

Accordingly, this chapter will now consider:

- land as the basis for taxation;
- land tax, stamp duty and rates;
- principles of valuation for tax on land and
- practice of valuation for tax on land.

7.2 Land as the basis for taxation

One of the fundamental characteristics of land is durability – land continues to exist in perpetuity and does not disappear, subject to such effects as coastal erosion and other climate impacts. Further, land is identifiable and measurable, lending easily to identification and registration of ownership and land area.

Such fundamental characteristics of land have made it a basis for taxation for millennia, including in biblical times:

> Jehoiakim paid Pharoah Neco the silver and gold he demanded. In order to do so, he taxed the land and exacted the silver and gold from the people of the land according to their assessments.
>
> (2 Kings 23:35; Biblica Inc, 2017)

and was one of the reasons for the compilation of the Domesday Book in Great Britain, completed in 1086.

Once a register of land ownership and area is created, the levying of tax on land is a relatively straight forward administrative process. Therefore, the taxation of land is capable of meeting the key criteria for a good tax as it is:

- efficient and provides a reliable, steady flow of income to Government;
- simple and easy to understand by taxpayers;
- equitable and fairly distributable across taxpayers; and
- transparent, with ease of clear assessment.

The ease with which tax may be levied on land is, however, a significant problem with State and Territory Governments and Councils each keen to increase the amount of revenue raised despite the negative flow on effects to the cost of housing and the cost of living to their electorates. Constraining such enthusiasm to tax has led to a wide range of differing control mechanisms being introduced by State and Territory Governments which add considerable complexity to the use of land value as a basis for taxation across Australia, with the ultimate control mechanism being the electorate itself.

7.3 Land tax, stamp duty and rates

In Australia, the three principal forms of taxation levied on land are:

- land tax;
- stamp duty and
- rates,

which are considered, respectively, below.

Hutley (2022) notes that there are differences in the way that these taxes on land are applied depending on:

- the State or Territory in which the property is located;
- the nature of the land use;
- whether the property is a principal place of residence or an investment;
- whether or not the purchaser is a resident Australian citizen; and
- whether or not the owner is in occupation or an absentee.

7.3.1 Land tax

In Australia, Reinhardt and Steel (2006) note that land taxes were first introduced by State Governments in the late nineteenth century after a long period of debate and blocking of such taxes by parliaments dominated by wealthy landowners. Land taxes were also introduced at the Federal level in 1910 as a form of wealth tax and as a means to break up large tracts of under-utilised land. In most States land was taxed at progressive rates, based on unimproved value, while the Federal land tax was introduced as a flat rate tax. It should be noted that, in the early 1900s, the currency of conversation in the community was the price of vacant building land sold for the erection of dwellings and for agricultural uses, rather than the sale price of houses.

As exemptions for primary production use of land increase, by the mid-twentieth century the bulk of land tax was levied on urban property rather than wealthy primary producers and large landholders. As urban markets developed and following the introduction of town planning, land values varied and land tax revenue became less stable. In 1952, land tax was abolished at a Federal level but continues to be applied at a State and Territory level. (Reinhardt and Steel, 2006)

Hutley (2022) notes that land tax is now levied by all State and Territory Governments, except the Northern Territory, against the value of a landholding. For strata title property, the land value for each strata lot is calculated on a proportional basis. In many jurisdictions there are categories of property that are exempt from land tax, often including primary places of residence, land for primary production, non-profit organisations, government, rooming houses, caravan parks and retirement villages.

The relevant statutes facilitating the various bases of value used for valuation for tax on land in the various States and Territories are as follows:

New South Wales	Valuation of Land Act 1916
Victoria	Valuation of Land Act 1960
Queensland	Land Valuation Act 2010
Western Australia	Valuation of Land Act 1978
South Australia	Valuation of Land Act 1971
Tasmania	Valuation of Land Act 2001
Northern Territory	Not levied
Australian Capital Territory	Rates Act 2004

The various bases of value used for valuation for land tax are summarised in Table 7.1.

Generally, land tax is assessed with regard to land value, land tax thresholds and rate in the dollar applied to land value above the threshold. Assuming a block of land with a value of $2 million upon which an investment property is situated in Sydney, the land tax assessment as at February 2025 would be calculated as follows:

Land value	$2,000,000
Less land tax general threshold	($1,075,000)
Assessable land value	$925,000
Land tax rate – $100 plus 1.6% of land value above the threshold	
Land tax liability	$14,900 pa

The amount of land tax paid by major investment property groups such as AREITs and property funds, major developers such as Meriton and Stockland as well as by private property investors in residential and non-residential investment property can be very significant. In order to minimise unrecoverable outgoings and so maximise returns on an investment property, it is in the interests of such groups to seek to reduce their land tax liability by as much as possible through the objection process and, if necessary, by appeal to the relevant Court or Tribunal.

7.3.2 Stamp duty

State and Territory Governments also use a form of land or property value as the basis for stamp duty assessment. Stamp duty is assessed at the time of sale and levied on the purchaser, with rates of stamp duty varying between jurisdictions and ownership types (for example, investment property may differ from primary place of residence).

Table 7.1 Bases of land value for land tax

State/territory	*Basis of value*
New South Wales	Land value
Victoria	Site value
Queensland	Unimproved value
Western Australia	Unimproved value
South Australia	Site value
Tasmania	Land value
Northern Territory	Not levied
Australian Capital Territory	Unimproved value

Source: Hutley (2022)

While some State and Territory Governments provide stamp duty concessions for first home buyers, some levy a stamp duty surcharge on foreign citizens buying residential property (Hutley, 2022).

As the value of residential property has increased across Australia, stamp duty on the transaction of residential property has become highly contentious and is often cited as a major contributor to the crisis in housing affordability. For example, in New South Wales, the stamp duty payable on the purchase of an existing residential property for owner occupation for $1 million is $39,529 as at February 2025.

7.3.3 Rates

At the local government level, Councils use a form of land or property value as the basis for levying rates. Rates are used to fund Council activities such as rubbish collection, parking, cemeteries, cultural facilities, community facilities and health services such as food inspection.

As rates are a highly visible expense to voters, State Governments often control how Councils set rates and increase rates with regular stories in the media of Councils trying to hike rates to balance their budget or having insufficient rate revenue to meet their expenses. For example, in New South Wales, overall rate revenue is increased on a three-yearly cycle to a level controlled by the Independent Pricing and Regulatory Tribunal (IPART) with Councils then determining the respective rate in the dollar and proportion of rates to be levied on residential property, commercial property and so on.

Rates may be ordinary rates or special rates and are usually applied to categories of land such as, in New South Wales under the *Local Government Act* 1993, residential, business, farming and mineral (Mangioni, 2006). Hutley (2022) notes that discounted rates and rebates are available in most jurisdictions, such as for pensioners and other concession card holders.

Rates are generally assessed on the basis of a rate in the dollar applied to the relevant basis of land value, sometimes involving thresholds or bands. Various bases of value used for the levying of rates are summarised in Table 7.2.

Table 7.2 Bases of land value for rates

State/Territory	*Basis of value*
New South Wales	Land value
Victoria	Site value
	Net annual value
	Capital improved value
Queensland	Unimproved value
Western Australia	Rural: unimproved value
	Non-rural: Gross rental value
South Australia	Land value
	Capital value
	Annual value
Tasmania	Land value
	Capital value
	Assessed annual value
Northern Territory	Unimproved capital value
	Annual value
	Improved capital value
Australian Capital Territory	Unimproved value

Source: Hutley (2022)

For example, the rates payable on a block of land with a value of $1 million upon which a house, being a principal place of residence, is situated in Sydney would be $1,410 pa as at February 2025.

7.4 Principles of valuation for tax on land

While there are many differences between the approaches to valuation for tax on land and rating purposes in each State and Territory, a common framework may be contended which is considered below and illustrated using New South Wales as an example. However, readers should carefully check the relevant legislation in the State or Territory where a valuation is undertaken for the specific State or Territory requirements.

7.4.1 Relevant statute

Section 7.3.1 listed the relevant statutes used for the valuation for tax on land in the various States and Territories.

While specifically defined in the respective statutes, valuation for tax on land is generally based on either the land value, unimproved value or site value. Hence the State and Territory Valuer Generals, their contract valuers and the valuers for landowners focus closely on the exact wording of the relevant statute in the assessment of value for tax on land.

7.4.1.1 Example – New South Wales

The relevant statute used for the valuation for tax on land in New South Wales is the *Valuation of Land Act* 1916.

Land tax is levied by the Chief Commissioner of Taxation and collected through the Office of State Revenue New South Wales. The onus is on the taxpayer to disclose property interests to the Office of State Revenue, rather than the onus being on the Office of State Revenue to discover.

Assessments of land value are provided annually by the Valuer General of New South Wales to the Office of State Revenue for the purposes of land tax and at least 3 yearly to Councils for the purpose of levying rates.

7.4.2 Basis of valuation

Table 7.1 listed the various bases of land value used for land tax purposes and Table 7.2 listed the various bases of land value used for the levying of rates.

While specifically defined in the respective statutes, the bases of land value are generally either the land value, unimproved value or site value. Hence the State and Territory Valuer Generals, their contract valuers and the valuers for landowners focus closely on the exact wording of the relevant statute in the assessment of value for tax on land.

7.4.2.1 Example – New South Wales

The basis of land value, for both land tax and the levying of rates, in New South Wales is land value as defined in the *Valuation of Land Act* 1916:

> **6A Land value**
>
> (1) The land value of land is the capital sum which the fee-simple of the land might be expected to realise if offered for sale on such reasonable terms and conditions as a

bona-fide seller would require, assuming that the improvements, if any, thereon or appertaining thereto, other than land improvements, and made or acquired by the owner or the owner's predecessor in title had not been made.

(2) Notwithstanding anything in subsection (1), in determining the land value of any land it shall be assumed that –

(a) the land may be used, or may continue to be used, for any purpose for which it was being used, or for which it could be used, at the date to which the valuation relates, and
(b) such improvements may be continued or made on the land as may be required in order to enable the land to continue to be so used,

but nothing in this subsection prevents regard being had, in determining that value, to any other purpose for which the land may be used on the assumption that the improvements, if any, other than land improvements, referred to in subsection (1) had not been made.

with land improvements defined in s4:

4 Definitions

(1) In this Act, unless inconsistent with the context or subject-matter –

Land improvements means –

(a) the clearing of land by the removal or thinning out of timber, scrub or other vegetable growths;
(b) the picking up and removal of stone;
(c) the improvement of soil fertility or the structure of soil;
(d) the restoration or improvement of land surface by excavation, filling, grading or levelling, not being works of irrigation or conservation;
(d1) without limiting paragraph (d), any excavation, filling, grading or levelling of land (otherwise than for the purpose of irrigation or conservation) that is associated with –

(i) the erection of any building or structure, or
(ii) the carrying out of any work, or
(iii) the operations of any mine or extractive industry,

(e) the reclamation of land by draining or filling together with any retaining walls or other works appurtenant to the reclamation and
(f) underground drains.

Conceptually, land value in New South Wales may be considered through the lens of the property market in 1916 when the majority of transactions were sales of blocks of land for the building of houses. Hence, the currency of conversation among residents of New South Wales in 1916 may be likely to have been how much somebody had paid for a block of land rather than how much somebody had paid for a house.

Despite the property market and the currency of conversation moving over the last century to focus on the capital value of houses rather than their land value, the basis of valuation remains land value. This is, in part, because it is the least worst alternative. On the one hand, market participants don't like it but do understand it, on the other hand value based on capital value raises electorally unpopular issues such as landowners being taxed for the improvements that they have made to their house.

The use of land value disregarding improvements and based on highest and best use does, however, encourage development of land for its highest and best use. For example, if a site is occupied by a 3 level office building but the highest and best use is for a 10 level office building, the disregard of existing improvements and adoption of highest and best use means that the owner is taxed on the potential of the site rather than its current under-developed state. Similarly, the adoption of highest and best use as the basis of value discourages land banking.

The definition of land value echoes the *Spencer* definition of market value, but 'assuming that the improvements' 'had not been made'. The land is, therefore, to be assumed to be a clear, levelled block ready for building that is unimproved except for removal of stones, scrub and so on. Following *Tooheys Limited v Valuer General* (1925) AC349 (PC), 'had not been made' should be considered to mean:

> Words could scarcely be clearer to show that the improvements were to be left entirely out of view. They are to be taken not only as non-existent, but as if they had never existed.
>
> (443)

Further, the land may be assumed to be used for its highest and best use (s6A(1)) or existing uses may be assumed to continue (s6A(2)) even if such uses no longer meet planning requirements such as being a more intensive use of land than currently permitted.

For example, assume two properties located adjacent on 500 sqm blocks of land zoned for single dwelling residential use. One of the properties is a single dwelling while the other property is a block of four flats built before the introduction of the current zoning instrument, therefore having an existing use right. The single dwelling will be valued under s6A(1) based on comparable sales of surrounding single dwelling houses. The block of flats will be valued under s6A(2) based on comparable sales of other unit development sites, which will probably be a higher value than the single dwelling value adjacent (Mangioni, 2006).

Section 7B of the Act addresses the valuation of land upon which a strata title property is constructed, with the valuation being of the underlying land not of the individual strata lots within the building. It is the role of the building's body corporate to assign a portion of land value to each lot which is usually based on unit entitlement.

For example, a strata scheme may have four lots with the following unit entitlements:

Lot 1	100
Lot 2	125
Lot 3	125
Lot 4	150
Total	500

If a value of $1,000,000 is assessed for the land owned by the strata scheme, this may be apportioned based on lot entitlement. Lot 1 would, therefore, be attributed 100/500 × $1,000,000 or $200,000 apportioned land value.

7.4.3 Frequency of valuations

Generally, valuations for tax on land will be undertaken on a recurrent cyclical basis with the date of valuation being known as the base date. There is commonly a tension between Governments' desire for regular updated valuations in order to maximise revenue, the cost of undertaking such valuations and the willingness of the community to pay taxes on land that may be the result of a short term upturn in the market which may soon be reversed by a downturn in the market.

Accordingly, Governments often seek a compromise by undertaking annual valuations (often using a mass appraisal methodology to reduce the cost of valuations – see Section 8.3.1 in Chapter 8) but then adopting an average of the last several years to smooth movements or adopting the annual valuation and varying the tax rate in the dollar applied to the valuation each year.

7.4.3.1 Example – New South Wales

The frequency of valuations in New South Wales is specified in the *Valuation of Land Act* 1916:

> **14A Valuer-General to ascertain land values**
>
> (1) The land value of each parcel of land in New South Wales, other than –
>
> (a) lands of the Crown, or
> (b) land that is within the Western Division and is not within the area of a rating or taxing authority,
>
> is to be ascertained each year.
>
> (2) The Valuer-General may at any time value any parcel of land, either on his or her own initiative or –
>
> (a) in the case of lands of the Crown, on the application of the public authority by or on whose behalf the land is held, or
> (b) in the case of land within the Western Division (including land referred to in paragraph (a)), on the application of the Secretary of the Department of Industry, or
> (c) in the case of land within the area of a rating or taxing authority (including land referred to in paragraph (a) or (b)), on the application of that authority.

Accordingly, annual valuations are undertaken by the Valuer General as at 1 July each year (s14(B)), being the base date, but not published until early the following year. Therefore, when taxpayers and rate payers receive their notice of valuation, it refers to a base date more than six months prior which, in periods of falling markets, may prompt an increase in objections.

To smooth the impact of annual valuation increases, the New South Wales Government uses the average of the last three years valuations for levying land tax and Councils are only permitted to adopt increased valuations on a three yearly basis, managing the interim period by variations in the rate in the dollar levied subject to rate pegging. It should be noted that s14(A)(2)(c) permits a Council to apply to the Valuer General for a valuation of any parcel of land at any time.

7.4.4 Valuation roll

Generally, the Valuer General will maintain a valuation roll or register which may be based on land ownership information provided by the relevant State or Territory Government land titles registry.

Such registers or rolls are usually available for valuers and landowners to access online and contain all relevant publicly available information about a land holding including tenure, caveats, easements, planning, heritage and so on, as well as current and historic data on the value of the land for taxation purposes.

7.4.4.1 Example – New South Wales

In New South Wales, the *Valuation of Land Act* 1916 specifies that:

> **14A Valuer-General to ascertain land values**
>
> (5) Any land value ascertained under this Act is to be entered in the Register of Land Values.

Accordingly, the Register of Land Values is a rich source of information for valuers undertaking assessments of the value of the land for taxation purposes.

7.4.5 Who undertakes the valuation?

In the early and mid-parts of the last century, the offices of the Valuer General in States and Territories were the largest employers of valuers in Australia undertaking the valuation of land for taxation purposes.

Then, in the later decades of the last century, there was a trend to outsourcing the provision of valuations to contract valuers with the significantly reduced staff of the Valuer General's offices focusing on quality control and policy issues.

As may be expected in the outsourcing/insourcing cycle, in the second decade of this century some Valuer Generals are revisiting the outsourced model and recruiting to undertake valuations internally once again.

7.4.5.1 Example – New South Wales

In New South Wales, the *Valuation of Land Act* 1916 specifies that:

> **13A Valuation service contracts**
>
> (1) The Valuer-General may negotiate and enter into valuation service contracts for the provision of valuation services to the Valuer-General.
>
> (2) Without limiting the terms and conditions of valuation service contracts, such contracts may regulate the manner in which valuation services are to be carried out. In particular, such contracts may contain provisions –
>
> (a) setting out the principles and methods according to which the valuation services are to be carried out,
>
> (b) establishing performance indicators to assist with assessing the effectiveness and efficiency of the valuation services that have been carried out,
>
> (c) prescribing the qualifications of persons (including contract valuers and employees or agents of contract valuers) who may exercise specified functions,
>
> (d) identifying persons (including contract valuers and employees or agents of contract valuers) who may or may not exercise specified functions,
>
> (e) regulating the manner in which specified functions are to be exercised,
>
> (f) imposing restrictions on the exercise of specified functions,
>
> (g) authorising contract valuers to exercise functions or discretions that expressly or impliedly belong to the Valuer-General,
>
> (h) setting out the obligations of contract valuers to provide assistance to the Valuer-General in dealing with objections under Part 3, defending appeals under Part 4 and exercising functions under Part 5.

(3) Contract valuers are not agents of and do not represent the Valuer-General, except where expressly provided by or under this or any other Act or by the terms of the relevant valuation service contract.

However, it should be noted that the contract valuer only makes a recommendation to the Valuer General and does not make the valuation – it is the Valuer General who makes the valuation:

13H Recommendations for valuations

(1) Any valuation for which the Valuer-General is required under section 13C to invite tenders for contested valuation service contracts must, and any other valuation under this Act may, be made by the Valuer-General on the recommendation of a contract valuer.

(5) The Valuer-General may request that a recommendation be revised by the contract valuer who prepared it.

7.4.6 Valuation notices

Generally, when a valuation is made, the Valuer General will issue a notice of valuation to the landholder. Such notice may be issued following a periodic revaluation or following an individual valuation such as after a subdivision has been registered.

7.4.6.1 Example – New South Wales

In New South Wales, the *Valuation of Land Act* 1916 specifies that:

29 Notice of valuations to owner

(1) On furnishing a valuation list to the council of a local government area, the Valuer-General must cause notice of each valuation contained in the list to be given to –

(a) the owner of the freehold estate in the land, and
(b) any lessee or occupier of the land who, under any Act, is liable to pay any rate or tax to a rating or taxing authority in respect of the land, and
(c) any lessee of the land under a written lease for a term exceeding 3 years who, under the lease, is liable to pay the whole or any part of any rate or tax to a rating or taxing authority in respect of the land, and
(d) any mortgagee in possession of the land.

As it is the owner of the freehold estate who is usually liable for tax on land, the requirement to issue a notice of valuation includes for both the purposes of rating and land tax. However, despite the valuation being made each year, a notice will only be issued to owners for rates purposes once every three years. The date of issue of the notice of valuation is important as it triggers the time period for an owner to object to the valuation assessment.

7.4.7 Objections

Generally, when a valuation notice is issued, the relevant statute will make provision for the land owner and, in certain cases, the lessee, to object to the valuation assessment within a given time frame.

The objection will usually be by completion of a specified form which may be provided online. Such a form will usually require the land to be identified, the grounds of objection to be stated, the objector's assessment of value to be provided and the evidence supporting the objection and assessment of value to be submitted.

The relevant statute will usually provide the bases upon which an objection may be made, such as the valuation is too low or too high. Objections on the ground that the land value is too low are often lodged by landowners who have leased the land with the rent under the lease linked to the Valuer General's assessment of the value of the land. In such circumstances, the lessee may also object that the land value is too high provided the lease comes within the provisions of the Act.

Objections on the basis that the land value is too high are much more common and should be supported by comparable sales evidence justifying the valuation contended by the owner. Percentage increases over the previous valuation or indices of price movements generated by such groups as Realestate.com or Domain.com are insufficient to ground an objection.

The objection and objector's contended valuation will be considered by the Valuer General either internally or outsourced to an independent valuer who will provide a valuation recommendation. Generally, it is not a case of proving that the valuation by the Valuer General is incorrect but rather of the objector proving that their valuation is correct and, by default, the Valuer General's is not.

There are generally three possible outcomes from a revaluation following objection:

- the land value contended by the objector is accepted and the objection is allowed;
- the land value contended by the objector is not accepted and the objection is disallowed; or
- the Valuer General's valuation is reduced, but not by the full amount sought by the objector,

with most objections ending at this point.

7.4.7.1 Example – New South Wales

In summary, the recipient of a valuation notice may lodge an objection to the valuation on the approved form within 60 days (or later if permitted by the Valuer General) on one of the bases for objection specified (too high, too low, factual error) and the Valuer General must consider the objection, either allow or disallow the objection and provide a notice of determination.

In New South Wales, the *Valuation of Land Act* 1916 specifies that:

> **29 Notice of valuations to owner**
>
> (3A) A person to whom the Valuer-General has given written notice under subsection (1) may lodge with the Valuer-General written objection to any such valuation.
>
> **33 Form of objection**
>
> An objection under this Part must be in a form approved by the Valuer-General.
>
> **34 Grounds of objection**
>
> (1) In relation to land the only grounds upon which objection may be taken under this Act are –
>
> (a) that the values assigned are too high or too low,
> (a1) that the area, dimensions or description of the land are not correctly stated,

(b) that the interests held by various persons in the land have not been correctly apportioned,
(c) that the apportionment of the valuations is not correct,
(d) that lands which should be included in one valuation have been valued separately,
(e) that lands which should be valued separately have been included in one valuation, and
(f) that the person named in the notice is not the lessee or owner of the land.

(2) In relation to a stratum the only grounds upon which objection may be taken under this Act are –

(a) that the values assigned are too high or too low,
(b) that the situation, description, or dimensions of the stratum are not correctly stated,
(c) that strata which should be included in one valuation have been valued separately,
(d) that strata which should be valued separately have been included in one valuation, and
(e) that the person named in the notice is not the lessee, occupier, or owner of the stratum.

(2A) In either case, an objection against the Valuer-General's refusal to determine an allowance or apportionment factor may be made on the ground that such an allowance or apportionment factor should have been determined.

(3) For the purposes of the application of this section to an objection under this Act to an allowance or an apportionment factor determined in respect of land or a stratum, a reference in this section to values is to be taken to include a reference to the amount of the allowance or the apportionment factor, as the case requires.

35 Time for lodging objection

(1) Except as provided by section 35A, an objection must be lodged with the Valuer-General, in accordance with the regulations, not later than 60 days after –

(a) the date of service of the notice of valuation under section 29, or
(b) in the case of a valuation for the purposes of the Land Tax Management Act 1956 or Property Tax (First Home Buyer Choice) Act 2022, the date of service of the relevant land tax assessment or property tax assessment under section 14 of the Taxation Administration Act 1996.

35A Objections lodged out of time

(1) The Valuer-General may permit a person to lodge an objection after the 60-day period.
(2) The person seeking to so lodge the objection must state fully and in detail, and in writing, the circumstances concerning and the reasons for the failure to lodge the objection within the 60-day period.
(3) The Valuer-General may grant permission unconditionally or subject to conditions or may refuse permission.
(4) The Valuer-General must give notice to the person of the Valuer-General's decision and include in the notice the reasons for refusing to grant permission or for imposing conditions on the permission.
(5) The notice is to be in a form approved by the Valuer-General.

35B Determination of objection

(1) The Valuer-General must consider an objection that has been duly made and either allow the objection or disallow the objection.
(2) If the Valuer-General delegates the functions conferred by this section, the delegate who considers the objection must be a different person from, and not subordinate to, the person who made the decision against which the objection is lodged.

35C Notice of determination

(1) The Valuer-General must give notice to the objector of the determination of the objection.
(2) The Valuer-General must, in the notice, give the reasons for disallowing an objection or for allowing an objection in part only.
(3) The notice is to be in a form approved by the Valuer-General.
(4) For the purposes of section 37, an objection is taken to have been disallowed if notice of the Valuer-General's determination of the objection has not been given within 90 days after the objection was lodged with the Valuer-General.

7.4.8 Appeal against objection decision

Generally, the relevant statutes will provide for an appeal process if the landowner is dissatisfied with the outcome of the objection with an appeal to be lodged within a defined timeframe. For example, in Victoria the right of appeal is to the Victorian Civil and Administrative Tribunal (VCAT), in Queensland to the Land Court of Queensland and in New South Wales to the Land and Environment Court.

It is important that the landowner has a valid basis for appeal and is not driven by emotion or ego. An error of law and/or an error of valuation will be required to ground an appeal and the landowner will need to produce evidence to the Court in support of their contention. Accordingly, the landowner would be well advised to retain a solicitor, a barrister, a valuer and other relevant consultants (planning, engineering, hydrology, heritage, etc.) as the Valuer General will be likely to have retained one of each. The provision of valuation evidence in litigation is a specialist area which is considered further in Section 8.5 in Chapter 8.

As with any form of litigation, appealing a land valuation objection may be an expensive exercise with Court fees for registering the appeal as well as professional fees for a solicitor, a barrister, a valuer and other experts. Some Courts or Tribunals will provide alternative dispute resolution opportunities, such as mediation or conciliation, as an early part of the appeal process. This provides the landowner with an opportunity to reach agreement with the Valuer General on a revised valuation without having to endure the expense and risk of a full hearing.

Generally, the decision of the Court or Tribunal is final and can only be appealed on a point of law with no appeal rights on points of valuation. Given the enormous number of valuation notices issued in States and Territories each year, the number of appeals to a Court or Tribunal is very low and the number that do not settle during alternative dispute resolution and so require a judicial decision is extremely low.

7.4.8.1 Example – New South Wales

In summary, a landowner dissatisfied with an objection assessment may appeal to the Land and Environment Court within 60 days after the issue of the notice of determination (or later if permitted by the Court) and may adopt different arguments on appeal to those adopted in the objection. The Court may confirm or revoke the Valuer General's valuation on objection, make an alternative valuation or send the matter back to the Valuer General to make a valuation determination in accordance with the Court's instructions.

In New South Wales, the *Valuation of Land Act* 1916 specifies that:

37 Right of appeal

(1) Any person entitled under Part 3 to object to a valuation may appeal to the Land and Environment Court if the person is dissatisfied with the Valuer-General's determination of any such objection to the valuation concerned (whether or not the person was the objector).
(2) An appeal may not be made on the ground that the objection is taken to have been disallowed, as referred to in section 35C (4), unless written notice of the objector's intention to appeal on that ground has been given to the Valuer-General at least 14 days before the appeal is made.
(3) No person or body has jurisdiction or power to conduct a review or hear an appeal in respect of the determination of an objection except as provided by this Part.

38 Time for appeal

(1) An appeal must be made not later than 60 days after the date of issue of the notice of the Valuer-General's determination of the objection.
(2) The Land and Environment Court may allow a person to appeal after the 60-day period.

39 Grounds of appeal

The appellant's and respondent's cases on an appeal are not limited to the grounds of the objection.

40 Powers of Land and Environment Court on appeal

(1) On an appeal, the Land and Environment Court may do any one or more of the following –
    (a) confirm or revoke the decision to which the appeal relates,
    (b) make a decision in place of the decision to which the appeal relates,
    (c) remit the matter to the Valuer-General for determination in accordance with the Court's finding or decision.
(2) On an appeal, the appellant has the onus of proving the appellant's case.

The level of objections and appeals in New South Wales each year is very low. The Valuer General reported 10,644 objections in 2022–23 (a general valuation year for rates and land tax purposes) and 4,903 in 2023–24 (a valuation year for land tax purposes only) to around 2.5 million valuations annually, of which only 12% and 30%, respectively, resulted in a change in land value (Valuer General, 2024).

Concerning appeals, in Class 3 which includes land valuation appeals, the NSW Land and Environment Court reported only 148 registrations in 2022 with 109 pre-trial disposals (such as agreement through mediation or withdrawal by the parties) and 18 matters disposed of by hearing and 176, 105 and 26, respectively, in 2023 (Land and Environment Court, 2024).

7.4.9 Valuation report

Concerning the layout of a valuation report prepared as part of the objection process and/or as part of a Court appeal, Mangioni (2006) sets out typical report contents as follows:

- description of property:
  - address of subject property;
  - legal description of subject property including encumbrances;
  - the date at which the land value is being challenged, generally known as the base date;
  - the contended land value;
  - description of the land (size, shape, dimensions, access, topography, etc.);
  - location of the property;
  - description of surrounding property;
  - proximity to amenities;
- description of the subject land – particularly focusing on shape, slope, views, impact of adjoining developments;
- description of the improvements:
  - construction;
  - age;
- town planning:
  - zoning and permitted use of land;
  - comment on whether existing use is the highest and best use;
  - height limits;
  - density permitted;
  - any heritage issues;
- grounds of objection;
- valuation method and approach – noting that the comparative method of valuation is usually preferred by the Courts and the use of an alternative method of valuation will require explanation;
- accumulation, analysis, adjustment and application of comparable sales and
- assessment of value.

Mangioni (2006) further suggests that the following be attached to the valuation report as annexures:

- subject property:
  - deposited plan;
  - certificate of title;
  - relevant planning certificates;
  - evidence of any contamination;
  - details of heritage issues;

- for each comparable sale used as sales evidence:
  - photograph;
  - title details and encumbrances;
  - deposited plan;
  - zoning details;
  - description of improvements;
  - description of land;
  - sale date (exchange and settlement);
  - sale price;
  - report of discussions with either selling agent or vendor, where possible and
  - explanation of the analysis and adjustments made.

7.4.10 Summary

While there are many differences between the approaches to valuation of land for land tax and rating purposes in each State and Territory, a common framework may be contended which was outlined and illustrated using New South Wales as an example. However, readers should carefully check the relevant legislation in the State or Territory where a valuation is undertaken for the specific State or Territory requirements.

7.5 Practice of valuation for tax on land – comparative sales method of valuation

As noted in Section 2.2 in Chapter 2, the Courts generally prefer the use of the comparative method of valuation following *Adams v Valuer General* [2014] NSWLEC 1005. For valuation for the taxation of land, the idiosyncratic features of a specific block of land compared to comparable sales require careful consideration such as the size, dimensions, topography and physical features and, particularly, aspect and views which may vary significantly between adjacent blocks of land.

For the purposes of illustration, the assessment of land value under the NSW *Valuation of Land Act* 1916 will be considered. Readers should note that, while the basis of value for the taxation of land may differ in other States and Territories, the comparative method of valuation is the preferred valuation method by the Courts.

For the purposes of applying the principles of *Adams v Valuer General* [2014] NSWLEC 1005 to the valuation of land under the *Valuation of Land Act* 1916, assume a subject property located at 12 Beach Street in a small northern New South Wales coastal settlement having a land area of 500 sqm and being of irregular topography.

The subject property is improved by a large modern dwelling with access to but elevated above the road, facing the beach with panoramic water views. However, the road is also elevated above beach level at this point such that access to the beach from the subject property requires pedestrians to cross the road and then either walk around the elevation or attempt a descent by climbing.

7.5.1 Accumulation step

The purpose of the accumulation step is to create a pool of relevant, potentially genuinely comparable sales from which information may be deduced concerning the value of the subject property.

Four potentially comparable sales may be identified, being:

20 Beach Street is comparable to the subject property in terms of location, vehicular access and views. However, the property is superior in terms of beach access but has significantly poorer topography and is significantly larger (1,000 sqm). Being improved by a very large modern dwelling, the property transacted for $2,500,000.

36 Beach Street is comparable to the subject property in terms of location, area (525 sqm) and views. However, the property is superior in terms of vehicular access, beach access and topography. Being improved by a modest beach house, the property transacted for $2,000,000.

38 Beach Street is comparable to the subject property in terms of location, area (475 sqm), topography and views. However, the property is superior in terms of vehicular access and beach access. Being improved by a modest beach house, the property transacted for $1,700,000.

36 Park Street is comparable to the subject property in terms of views. However, the property is superior in terms of location, topography, vehicular access and beach access but is larger (750sqm) than the subject property. Being improved by a large modern dwelling, the property transacted for $2,750,000.

7.5.2 Analysis step

The purpose of the analysis step is to provide a common basis of measurement by seeking to convert all potentially comparable sales to a common basis of expression such as a unitary rate (rate per square metre, rate per hectare, etc.), improved or unimproved (through allowance for the absence or existence of improvements) and so forth.

Accordingly, each sale requires analysis to deduct the value of the house from the sale price and then express the resultant land value as a rate per square metre as shown in Table 7.3.

7.5.3 Adjustment step

The purpose of the adjustment step is to acknowledge that no two properties are ever identical through seeking to convert each potentially comparable sale to a hypothetical expression of value as a unitary rate in the context of the subject property through the reflection of differences (such as size, location, use, date and so forth). Such adjustment may be expressed through percentages, arrows, ticks or other comparative expressions provided that same are used consistently

Accordingly, each sale requires adjustment to reflect differences with the subject sale as shown in Table 7.4.

Table 7.3 Analysis step

Address	*Sale price*	*Less house*	*Analysed price*	*Land area*	*Analysed rate psm*
20 Beach St	$2,500,000	$450,000	$2,050,000	1,000 sqm	$2,050
36 Beach St	$2,000,000	$250,000	$1,750,000	525 sqm	$3,333
38 Beach St	$1,700,000	$225,000	$1,475,000	475 sqm	$3,105
36 Park St	$2,750,000	$375,000	$2,375,000	750 sqm	$3,166

Source: Author

Table 7.4 Adjustment step

	20 Beach St	*36 Beach St*	*38 Beach St*	*36 Park St*
Site area	–10%	0%	0%	–15%
Location	0%	0%	0%	–10%
Topography	+25%	–5%	0%	–5%
Views	0%	0%	0%	0%
Vehicular access	0%	–10%	–10%	–10%
Beach access	–10%	–10%	–10%	–10%
Total	+5%	–25%	–20%	–50%

Address	*Analysed price*	*Analysed rate psm*	*Adjustment*	*Adjusted rate psm*
20 Beach St	$2,050,000	$2,050	+5%	$2,152
36 Beach St	$1,750,000	$3,333	–25%	$2,500
38 Beach St	$1,475,000	$3,105	–20%	$2,484
36 Park St	$2,375,000	$3,166	–50%	$1,583

Source: Author

7.5.4 Application step

The purpose of the application step is to determine the value of the subject property through the consideration of the relevance (such as being limited, indirect or direct) of the unitary rate derived from the adjusted comparable sales relative to the subject property.

Reflecting the number of adjustments and the quantum of adjustment required, 36 Park Street is considered to be an unsafe comparable sale to use, with the balance of comparable sales considered directly relevant.

Applying the range of directly relevant adjusted rates psm ($2,152–$2,500 psm) to the area of the subject property of 500 sqm provides a land value range of $1,076,000-$1,250,000. As the sample of directly relevant adjusted rates psm are skewed towards the upper end of the range, a land value of $1,200,000 ($2,400 psm) is assessed for the subject property for the purposes of tax on land.

7.6 Summary and conclusions

Focusing on the key issues from a valuer's perspective, Chapters 1 and 2 introduced the concepts and principles of statutory valuation, with statutory valuation for acquisition considered in Chapters 3, 4, 5 and 6. Statutory valuation for rating and taxing is considered in Chapters 7 and 8, with regulatory valuation then considered in Chapters 9 and 10.

Chapter 1 introduced the principles of statutory valuation, including the concepts of market value, liberal/conservative estimates of value, highest and best use, potentiality and confirmation of a foresight. This provided the essential foundation principles of statutory valuation that are then applied in the following chapters on statutory valuation.

Chapter 2 considered the practice of statutory valuation including valuation methods with an emphasis on the comparative method of valuation, together with statutory valuation conventions such as the before and after method and the piecemeal method. This provided an understanding of the valuation methods and conventions that may be preferred for application in the following chapters on statutory valuation.

Statutory valuation for acquisition is considered in chapters 3, 4, 5 and 6. Chapter 3 introduced the principles of statutory valuation for acquisition, including the statutory framework and process, with a focus on the concept of disregard of the scheme in compensation assessment and the distinction between concepts of value and worth. This provided an insight into concepts of fairness and equity in statutory valuation in the context of the disregard of compulsion.

Chapter 4 explored the market value bases of compensation in acquisition for which statutory valuation may be required, including market value of the interest and the related market value compensation concepts of severance and injurious affection. This provided a broad understanding of the heads of compensation for acquisition on a value basis that may arise in statutory valuation.

Chapter 5 considered non-market value forms of compensation including special value which was noted to rarely arise, payment for solace arising from loss of home which is heavily dependent on the exact wording of the facilitating statute and the concept of disturbance for costs and losses suffered as a result of the acquisition. This provided a broad understanding of the heads of compensation for acquisition that are not, necessarily, on a value basis but that may arise in statutory valuation.

Chapter 6 concluded the chapters considering statutory valuation for acquisition by examining compensation for other interests such as easements and mortgages, mineral rights and profits à prendre, strata title and substrata, businesses and the emerging areas of native title and of indigenous cultural loss. This provided an appreciation of the diversity of interests that may be subject to acquisition and so may require valuation for the assessment of compensation.

Statutory valuation for rating and taxing is considered in Chapters 7 and 8. Chapter 7 provides a general understanding of the often arcane and artificial basis of statutory valuation upon which Government collects very significant amounts of taxation revenue, introducing the principles of durability, identifiability and measurability which make land an ideal basis for taxation. Taxation through land tax, stamp duty and rates were considered with the myriad of variations to the underlying principles arising from State and Territory legislation outlined. The electorally visible nature of taxes on land, especially stamp duty and rates, was noted together with the significant amount of revenue raised, making taxes on land highly politically sensitive.

While the process of valuation for tax on land varies marginally between States and Territories, the process in New South Wales was considered to illustrate the key elements. Valuation basis, frequency, roll, insourcing/outsourcing, notices, objections and appeals were considered, noting the extremely low level of objections with the low level of changes in valuations suggesting that the process generally operates effectively and fairly for the benefit of taxpayers.

Given the importance of the comparative sales method of valuation for statutory valuation generally and for the valuation of land for the purposes of taxation particularly, an example of the accumulation, analysis, adjustment and application process was provided for the valuation of land which identified the importance of idiosyncratic features such as topography, views and beach access to the assessment of value for land taxation purposes.

Chapter 8 explores the practice of statutory valuation for rating and taxing through exemptions, concessions and allowances that have evolved over the last century, together with contemporary issues such as mass appraisal, automated valuation models and artificial intelligence, the impact of major disasters on valuation for rating and taxing and the litigation process including the valuer as an expert witness. This provides an insight into the continued evolution of statutory valuation to address issues as they arise over time, highlighting the importance of a principles-based approach to statutory valuation

Regulatory valuation, considered in the context of the preparation of valuations for inclusion in financial reporting, is considered in Chapters 9 and 10. Chapter 9 explores the principles of regulatory valuation with regard to the inter-related roles of international accounting standards, international financial reporting standards and international valuation standards, focusing on the primacy of fair value in international financial reporting standards. This provides an understanding of the fundamental conceptual differences and bases between statutory valuation and regulatory valuation.

Finally, Chapter 10 concludes regulatory valuation by exploring the practice of valuation approaches and methods relevant for regulatory valuation such as the income approach and the cost approach and the various valuation methods there within. This provides an appreciation of the wider range of valuation approaches and methods available to the valuer for regulatory valuation reflecting that, unlike statutory valuation, there is no requirement for defence in Court (negligence claims excepted).

References

Biblica Inc (2017) *Holy Bible – new international version*, Zondervan, Grand Rapids

Hutley, N (2022) Policy principles, in D Parker (Ed) *Principles and practice of property valuation in Australia*, Routledge, Abingdon

Land and Environment Court (2024) *The Land and Environment Court of NSW annual review 2023*, State of New South Wales, Sydney

Mangioni, V (2006) *Land tax in Australia*, Australian Property Publications, Bondi

Parker, D (Ed) (2022) *Principles and practice of property valuation in Australia*, Routledge, Abingdon

Parker, D (2024) *Introduction to property valuation in Australia*, Routledge, Abingdon

Reinhardt, S and Steel, L (2006) *A brief history of Australia's tax system*, paper presented at the 22nd APEC Finance Ministers' Technical Working Group Meeting in Khanh Hoa, Vietnam, 15 June

Valuer General (2024) *Yearly insights 2023–24*, Valuer General, Sydney

8 Rating and land tax

Practice

8.1 Introduction

This book is an introduction to two very different valuation worlds in Australia – the complex and often arcane national realm of statutory valuation and the dynamic global empire of regulatory valuation. Statutory valuation is different from regulatory valuation, with the former being based on relevant applicable statutes and being capable of challenge on appeal in Court while the latter is based on relevant applicable regulations and incapable of challenge in Court (negligence claims excepted).

As an introductory textbook on statutory and regulatory valuation, this book is a companion to the Australian valuation textbooks, *Introduction to Property Valuation in Australia* (Parker, 2024) and *Principles and Practice of Property Valuation in Australia* (Parker, 2022), which provide a wider analysis of key principles underlying property valuation and current techniques and issues in the practice of property valuation for major sectors of the Australian property market.

Statutory valuation in Australia is most commonly encountered in valuations for rating, taxing and acquisition but may arise wherever a statute prescribes a basis of or approach to valuation. Regulatory valuation is most commonly encountered in valuations for financial statements but may arise wherever a regulation prescribes a basis of or approach to valuation.

Rating and taxing of land raise significant amounts of revenue for State Governments and Councils in Australia, based on statutory valuation. For example, the NSW 2025–2026 Budget Statement estimates land tax revenue for 2024–2025 to be $8.2 billion with the City of Brisbane Annual Plan and Budget 2024–2025 estimating proposed rates and utility charges to be $1.51 billion.

Acquisition of land, property and businesses commonly arise for major infrastructure projects and may be by agreement or by compulsion, based on statutory valuation. For example, such major infrastructure projects in Australia currently include the multi-billion-dollar Sydney Metro, Melbourne Suburban Rail Loop and Melbourne to Brisbane Inland Rail projects.

Valuation of land and property holdings based on regulatory valuation arises for all reporting entities in Australia that publish financial statements upon which it is reasonable to expect that users will rely for decision making. For example, at June 2025 there are over 2,000 reporting entities listed on the Australian Stock Exchange which each publish financial statements, with the financial statements for Goodman Group, as at 30 June 2024, showing the fair value of total property assets to be $20.2 billion, based on regulatory valuation.

Focusing on the key issues from a valuer's perspective, the concepts and principles of statutory valuation are introduced in Chapters 1 and 2, with statutory valuation for acquisition

DOI: 10.1201/9781032616957-8

considered in Chapters 3, 4, 5 and 6 and statutory valuation for rating and taxing considered in Chapters 7 and 8. Regulatory valuation is then considered in Chapters 9 and 10.

Chapter 1 introduced the principles of value in statutory valuation, including the concepts of market value, liberal/conservative estimates of value, highest and best use, potentiality and confirmation of a foresight. This provided the essential foundation principles of statutory valuation that are then applied in the following chapters on statutory valuation.

Chapter 2 considered the practice of statutory valuation including valuation methods for statutory valuation with an emphasis on the comparative method of valuation, together with statutory valuation conventions such as the before and after method and the piecemeal method. This provided an understanding of the valuation methods and conventions that may be preferred in the following chapters on statutory valuation.

Statutory valuation for land acquisition is considered in Chapters 3, 4, 5 and 6, supported by an evolving example in Appendix 1 of each chapter. Chapter 3 introduced the principles of statutory valuation for land acquisition, including the statutory framework and process, with a focus on the concept of the disregard of the scheme in compensation assessment and the distinction between concepts of value and worth. This provided an insight into concepts of fairness and equity in statutory valuation in the context of the requirement to disregard the compulsory nature of the acquisition.

Chapter 4 explored the market value bases of compensation in acquisition for which statutory valuation may be required, including market value of the interest and the related market value compensation concepts of severance and injurious affection. This provided a broad understanding of the heads of compensation for acquisition on a value basis that may arise in statutory valuation.

Chapter 5 considered non-market value forms of compensation including special value which was noted to rarely arise, payment for solace arising from loss of the primary place of residence and the concept of disturbance for costs and losses suffered as a result of the acquisition. This provided a broad understanding of the heads of compensation for acquisition that are not, necessarily, on a value basis but which may arise in statutory valuation.

Chapter 6 concluded the chapters considering statutory valuation for acquisition by examining compensation for other interests such as easements and mortgages, mineral rights and profits à prendre, strata title and substrata, businesses and the emerging statutory valuation areas of native title and indigenous cultural loss. This provided an appreciation of the diversity of interests that may be subject to acquisition and so may require valuation for the assessment of compensation.

Statutory valuation for rating and taxing is considered in Chapters 7 and 8. Chapter 7 introduced the principles of land as the basis for taxation through land tax, stamp duty and rates, considering the general principles of the valuation of land for taxation using the New South Wales statutory framework as an example. This provided a general understanding of the often arcane and artificial basis of statutory valuation upon which Government collects very significant amounts of taxation revenue.

This chapter explores the practice of statutory valuation for rating and taxing through exemptions, concessions and allowances that have evolved over the last century, together with contemporary issues such as mass appraisal, automated valuation models and artificial intelligence, the impact of major disasters on valuation for rating and taxing and the litigation process, including the valuer as an expert witness. This provides an insight into the continued evolution of statutory valuation to address issues as they arise over time, highlighting the importance of a principles-based approach to statutory valuation.

Regulatory valuation, considered in the context of the preparation of valuations for inclusion in financial reporting, is considered in Chapters 9 and 10. Chapter 9 explores the principles of regulatory valuation with regard to the inter-related roles of international accounting standards, international financial reporting standards and international valuation standards, focusing on the primacy of fair value in international financial reporting standards. This provides an understanding of the fundamental conceptual differences and bases between statutory valuation and regulatory valuation.

Finally, Chapter 10 concludes regulatory valuation by exploring the practice of valuation approaches and methods relevant for regulatory valuation such as the income approach and the cost approach and the various valuation methods there within. This provides an appreciation of the wider range of valuation approaches and methods available to the valuer for regulatory valuation reflecting that, unlike statutory valuation, there is no requirement for defence in Court (negligence claims excepted).

Accordingly, this chapter will now consider:

- exemptions, concessions and allowances;
- mass appraisal, automated valuation models and artificial intelligence;
- impact of major disasters and
- litigation and expert witness.

8.2 Exemptions, concessions and allowances

As taxes on land are amongst the oldest taxes in Australia, they have evolved significantly over the last century in the various States and Territories following economic and political developments. While there is some commonality in exemptions and concessions between States and Territories, there are considerable differences in allowances.

8.2.1 Exemptions and concessions

The common exemptions from land tax in all States and Territories are for:

- principal place of residence;
- land used for primary production and
- government owned land.

Further exemptions and concessions are available in various States and Territories as follows:

8.2.1.1 New South Wales

Exemptions and concessions include:

Boarding houses
Low-cost accommodation
Residential and caravan park used as a home for a community of senior or retired persons
Non-profit organisations
Retirement villages, aged care establishments and nursing homes
Child care centres
Societies, clubs and associations

Source: https://www.revenue.nsw.gov.au/taxes-duties-levies-royalties/land-tax/exemptions-and-concessions/other-exemptions, accessed 10 March 2025

8.2.1.2 Victoria

Exemptions and concessions include:

Charities
Rooming houses
Friendly societies
Social housing and emergency housing
Housing provided for the relief of poverty
Sporting, outdoor recreation or outdoor cultural activities land owned by not-for-profit organisations
Land leased for outdoor sporting, recreational or cultural activities by members of the public
Health centres and services (not commercially operated businesses)
Residential care facilities and supported residential services
Residential services for people with disabilities
Retirement villages
Caravan parks
Mines
Agricultural shows and farm field machinery days
Not-for-profit clubs (excluding horse and harness racing clubs and private gender-exclusive and gender-restrictive clubs)
Land protected by a conservation covenant with Trust for Nature
with a concessional tax rate available for clubs promoting horse and harness racing and a land tax concession for build-to-rent developments.

Source: https://www.sro.vic.gov.au/land-tax/land-tax-exemptions, accessed 10 March 2025

8.2.1.3 Queensland

Exemptions and concessions include:

Charitable institutions
Moveable dwelling (caravan) parks
Aged care facilities
Retirement villages
Supported accommodation
Society, club or association

Source: https://qro.qld.gov.au/land-tax/relief/exemptions/, accessed 10 March 2025

8.2.1.4 South Australia

Exemptions and concessions include:

Affordable community housing
Affordable housing

Where the homeowner has moved into residential care
Where the homeowner has died
Where you are renovating or rebuilding your residential home
Where your residential home has been destroyed or rendered uninhabitable
Supported residential facilities and aged care facilities
Retirement village
Retired persons' relocatable home park
Association or not-for-profit organisation
Purchased as an off-the-plan apartment
Where subject to a heritage agreement under the Native Vegetation Act 1991
A build-to-rent land tax concession.

Source: https://www.revenuesa.sa.gov.au/land-tax/ltexemptions, accessed 10 March 2025

8.2.1.5 Tasmania

Exemptions and concessions include:

New home made available for long term rental
Non-profit and related organisations
Trustee of a special disability trust where land used as the principal residence of a beneficiary of a trust
Properties converted from short term visitor accommodation to a long term lease
Two residences owned in transitional circumstances
New home builders rebate, a home business concession and a special rate of tax for eligible clubs.

Source:https://www.sro.tas.gov.au/land-tax/exemptions-and-rebates/special-rate-of-tax, accessed 10 March 2025

8.2.1.6 Western Australia

Exemptions and concessions include:

Moving between homes
Building a new home
Moving to a new or renovated home
Deceased estate
Owner in care
Right to residue under a will
Property used by a related disabled person
Trust property used by a disabled beneficiary
Inner city living rebate
Religious institution
Educational institution
Aged care facility
Non-profit or sports association

Caravan parks
Retirement villages
Build-to-rent
Charitable exemptions
Subdivided residential land
Land used for public or religious hospital
Land held as mining tenements
Land dedicated to zoological, agricultural or other public purposes
Land owned by a veteran's surviving partner or mother
Land held under an approved conservation covenant
Land vacated for sale by mortgagee

Source: https://www.wa.gov.au/service/financial-management/taxation-and-duty/apply-land-tax-exemption, accessed 10 March 2025

8.2.1.7 Australian Capital Territory

Exemptions and concessions include:

Up to 3 months after settlement to move into principal place of residence, provided it is not rented
Up to 1 full quarter after moving out of principal place of residence, provided it is not rented
Deceased estate up to 2 years after owner's death or until transferred to a beneficiary or becomes rented, which may be extended on application
Homes occupied by a person with a life or term interest under a will
Up to 2 years for former principal place of residence if you can no longer live independently and now live in a hospital, hospice, mental health or community care facility, nursing home or with a carer
Home that is unfit for occupation if under construction, being significantly renovated or has been damaged
Home that is occupied by a person rent free or for a nominal rent
Rural land
Land owned by the Housing Commissioner
Transfers to entities for community housing
Retirement villages
Nursing homes
Religious accommodation
Land for prescribed purposes
Development leases (broad acre subdivisions)
Land owned by a not-for-profit housing corporation
Affordable community housing
Compassionate reasons

Source: https://www.revenue.act.gov.au/land-tax/exemptions, accessed 10 March 2025

8.2.1.8 Northern Territory

There is no land tax in the Northern Territory.

8.2.1.9 Summary

Over the last century, while the principal place of residence and primary production land have remained constant, a wide range of exemptions and concessions between the States and Territories has developed and continues to evolve.

While including aged care facilities, nursing homes and retirement villages reflects the structural demographic shift in Australia's ageing population, the inclusion of low-cost accommodation, social/affordable/community housing, emergency housing and build to rent reflects the effects of Australia's housing shortage.

Often reflecting political pressures and imperatives, exemptions and concessions change regularly and readers should check the relevant State or Territory website for the most recent list.

While there is some commonality in exemptions and concessions between States and Territories, allowances differ significantly between States and Territories as considered in Section 8.2.2.

8.2.1.10 Land Tax Management Act 1956 (NSW)

Further, it should be noted that, in New South Wales, s9A and s62I-s62N of the *Land Tax Management Act* 1956 also provide a concession for unutilised land value:

> s9A Concession for unutilized land value
>
> (2) For the purpose of assessing land tax, the land value of the land is to be reduced by the unutilised value allowance.
>
> s62J Land that is eligible to have unutilised value ascertained
>
> (1) Land is eligible to have an unutilised value allowance ascertained for its land value as at 1 July in a year if it satisfies the description in either of the following paragraphs as at midnight on 30 June in that year –
>
> (a) a parcel of land used or occupied solely as the site of a single dwelling-house and which is, under an environmental planning instrument, zoned or otherwise designated for use for the purposes of industry, commerce or the erection of residential flat buildings,
>
> (b) a parcel of land (which may comprise one or more lots or portions in a current plan within the meaning of the Conveyancing Act 1919) used or occupied solely as the site of a single dwelling-house and which is, under an environmental planning instrument, zoned or otherwise designated, so as to permit its subdivision for residential purposes.
>
> 62L How unutilised value allowance is ascertained
>
> (1) The unutilised value allowance for a land value is the amount calculated by deducting from the land value of the land the value that the land would have if it could be used only as the site of a single dwelling-house.
>
> (2) However, no account is to be taken of any portion of the land which is in excess of that which is reasonably necessary to be occupied or used in conjunction with the single dwelling-house.

This concession primarily applies to single dwelling residential property situated on land zoned for commercial or industrial uses or land that may be used for residential flat buildings, with the concession relating to the value of the unutilised portion of the land which is set aside.

8.2.2 Allowances

Allowances refer to the various ways in which different States and Territories make provision for land tax relief for different situations, usually within the relevant legislation. As it is impractical to list all of the allowances provided by all of the States and Territories, those allowances provided in New South Wales under the *Valuation of Land Act* 1916 will be used as an example.

Under the *Valuation of Land Act* 1916, allowances are addressed in the following sections:

- Division 3 Allowances for profitable expenditure;
- Division 4 Allowances for subdivision;
- Division 5 Apportionment factors for mixed development land and
- Division 5A Apportionment factors for mixed use land,

together with provisions on how specific types of property are to be valued, such as heritage property and business incidental to principal place of residence.

8.2.2.1 Division 3 – Allowances for profitable expenditure

Profitable expenditure is considered in s14L-s14R of the *Valuation of Land Act* 1916, which address an allowance for expenditure incurred in relation to the land with the allowance lasting for a period of 15 years or expiring if the property is sold within that period.

Such expenditure may be for land improvements on or to the land as defined in s4 of the *Valuation of Land Act* 1916 such as excavation, filling or clearing. Expenditure may also include visible and effective improvements off the land such as roadworks, electricity supply, water and drainage improvements. Although not on the land such improvements have been provided for the benefit of the land and have added value to the land.

The allowance must not exceed the cost of the improvements as at the date at which the allowance is being determined and may be either contended to the Valuer General by the landowner with relevant supporting evidence or determined by the Valuer General (Mangioni, 2006).

For example, a landowner may spend $100,000 on excavating, filling, grading and levelling their block of land. Depending on the change in value, each of the following scenarios may arise:

Scenario 1	
Value of site before works	$500,000
Value of site after works	$600,000
Cost of works	$100,000
Value added by the works	$100,000
Allowance	$100,000
Scenario 2	
Value of site before works	$500,000
Value of site after works	$650,000
Cost of works	$100,000
Value added by the works	$150,000
Allowance	$100,000 as it cannot exceed the cost of works.

(*Continued*)

(Continued)

Scenario 3	
Value of site before works	$500,000
Value of site after works	$550,000
Cost of works	$100,000
Value added by the works	$50,000
Allowance	$50,000 as the works only added $50,000 in value despite costing $100,000

It should be noted that the land value is not reduced but remains as assessed, with the allowance provided separately. The taxable value is reduced by deducting the allowance from the land value.

8.2.2.2 Division 4 – Allowances for subdivision

Allowance for subdivision is considered in s14S–s14W of the *Valuation of Land Act* 1916, which address a discount to the value of the individual lots in a subdivision had they been sold separately and the value of the lots had they been sold to one person – effectively a discount for a sale in one line (Mangioni, 2006).

A subdivider, as defined in s14S, must apply for an allowance prior to the registration of the plan of subdivision on the basis of owning all the land comprising the lots in the plan. If the land qualifies for an allowance for subdivision, the Valuer General must ascertain the allowance as a discount from the sale price in accordance with s14U(2):

> (2) The discount from sale price, in relation to lots in a deposited plan, means the amount (if any) that in the opinion of the Valuer-General is the difference between –
>
> (a) the total of the land values of the lots had they been sold separately, and
> (b) the total of the land values of the lots had they been sold to one person.

For example, a landowner who is a subdivider may hold a subdivision of 10 lots that will each sell for $100,000 yielding $1,000,000 in total. However, if all 10 lots were sold in one line, the sale price may be $900,000. Accordingly, the Valuer General may ascertain a subdivision allowance of $100,000 or 10% for each lot. While an allowance of 10% may be commonly adopted, it should be supported by examples of discounts evidenced by sales of land in one line where available.

8.2.2.3 Division 5 – Apportionment factors for mixed development land and Division 5A – Apportionment factors for mixed use land

Apportionment factors are considered in the *Valuation of Land Act* 1916 as follows:

Division 5	s14X-s14BB Apportionment factors for mixed development land; and
Division 5A	s14BBA-s14BBE Apportionment factors for mixed use land.

Mixed development land apportionment factors apply where there is a combination of uses of land, for example a shop and a residence, and the landowner applies for an allowance. The apportionment factor is expressed as a percentage, being the rental value of that part of the land that is occupied or is used for non-residential purposes relative to the rental value of the mixed development land as a whole (Mangioni, 2006).

Mixed use land apportionment factors apply where there is a residential use and non-residential use within the one parcel of land, being where there are residential and non-residential uses within a parcel that don't fit within the definition of mixed development land.

Mixed development land apportionment factors are used by Councils to allow for separate rating categories for residential and business lands, with Councils generally not using mixed use land apportionment factors. Mixed development land apportionment factors and mixed use land apportionment factors are used by State Revenue to separate principal place of residence exemptions on mixed use land.

For example, a shop with residence above may have a land value of $1,500,000. The shop may have a rental value of $100,000 pa and the residence may have a rental value of $50,000 pa. The mixed development apportionment factor would be determined as follows:

Shop rental	$100,000 pa
Residence rental	$50,000 pa
Total rental	$150,000 pa

The mixed development apportionment factor is calculated as:

$$\frac{\$100{,}000}{\$150{,}000} = 66.67\%$$

The Land Value is then apportioned as:

Shop proportion	$1,500,000 × 66.67%	= $1,000,000
Residence proportion	$1,500,000 – $ 1,000,000	= $500,000

8.2.2.4 Provisions on how specific types of property are to be valued

The *Valuation of Land Act* 1916 also addresses how the following specific types of property are to be valued:

- s14C Valuation of land in the Western Division;
- s14D Protected archaeological areas, wildlife districts, wildlife refuges and game reserves;
- s14E Community schemes, neighbourhood schemes and certain strata schemes;
- s14F Valuation of mines and minerals;
- s14G Valuation subject to heritage restrictions under an Environmental Planning Instrument;
- s14H Valuing rent-controlled land; and
- s14I Valuing Crown lease restricted land.

The most controversial and complex of the above is s14G which addresses the valuation of property subject to heritage restrictions under an Environmental Planning Instrument:

14G Valuation subject to heritage restrictions under EPI

(1) Land that is heritage restricted on the date by reference to which its land value is to be determined is to have its land value determined on the basis of the following assumptions –

(a) that the land may be used only for the purpose, if any, for which it was used when the value is determined,

(b) that all improvements on that land when the value is determined may be continued and maintained in order that the use of that land as referred to in paragraph (a) may be continued, (b1) that all improvements referred to in paragraph (b) on that land are new (without any deduction being made because of their actual condition),

(c) that no improvements, other than those referred to in paragraph (b), may be made to or on that land,

(d) that the cost of construction of improvements on that land has no effect on its land value, with the result that there is to be no reduction in land value because of any difference between the cost of construction of the improvements referred to in paragraph (b) as new improvements and the cost of construction of other improvements used as a basis for comparison in the determination of land value.

In essence, in order to provide land tax relief for owners of heritage properties for whom redevelopment is constrained, s14G seeks to provide an allowance for the impact of heritage on the value of the property. To date, such allowance has been determined with regard to relative rental values.

In its simplest form for a heritage property used for offices, the allowance may be calculated by determining the rental income from the property and comparing that to the rental income had the heritage listing not been in place. As such, the allowance may differ significantly between different heritage properties depending on the characteristics of the specific property.

For an office property, the rental income from the property compared to the rental income had the heritage listing not been in place will usually be influenced by the floor space ratio, the market rental and the level of outgoings. For example, consider a heritage office building in the CBD with a sandstone façade with wide staircases but no lifts nor air conditioning:

- assume the building has a gross floor area of 10,000 sqm and a net lettable area of 7,500 sqm giving a building efficiency of 75%, reflecting the extent of the staircases, which is considered to be in line with other heritage office properties and therefore no further adjustment is required;
- assume the permitted FSR is 12.5:1 with the subject property showing an FSR of 10:1;
- assume the rental level in the subject property is $750 psm pa while the market rent of surrounding non-listed buildings is $1,000 psm pa;
- assume that the land value without allowance for heritage has been assessed as $6 million.

The heritage allowance may be assessed as follows:

- impact of FSR – the subject property is 20% below the permitted FSR due to the heritage listing; and
- market rent impact – the subject property rent is 25% below the market rent due to the heritage listing.

Therefore, the sum of the FSR impact (20%) and market rent impact (25%) is 45% providing a 45% allowance for heritage of $2,700,000 resulting in a revised land value of $3,300,000.

While this is a simple example to illustrate the principles, more complex valuation examples may be found in *Commonwealth Custodial Services Ltd v Valuer General* [2008] NSWLEC

310, *Oriental Bar Pty Limited v Valuer-General* [2015] NSWLEC 59 and *Valuer-General of New South Wales v Oriental Bar Pty Ltd* [2016] NSWCA 48.

8.3 Mass appraisal, automated valuation models and artificial intelligence

Historically, the practice of statutory valuation saw a valuer from the Valuer General's office individually value each property in a State or Territory. While this approach was time consuming and expensive, it was contended to ensure that each valuation had appropriate regard to the specific characteristics of the individual property being valued.

Towards the end of the last century, some Valuer Generals not only outsourced the valuation process to private contractors but also moved away from valuing each property individually to valuing groups of properties with similar characteristics, a process commonly referred to as mass appraisal. Over time, mass appraisal has become accepted by the community and Government has benefitted from significant cost savings in the valuation process.

The valuation of property using computer-based valuation models based on the characteristics of the property relative to comparable sales, known as automated valuation models, have been taught in the valuation courses of the world's leading universities for over 40 years though are yet to replace the role of valuers in Australia.

However, it is in the confluence of large amounts of data, large amounts of computing power and artificial intelligence that the greatest challenge to the role of the valuer may emerge in this century. While this may provide major cost savings and a much faster valuation process, how idiosyncratic features of individual properties may be reflected in an assessment of value remains a significant challenge.

8.3.1 Mass appraisal

Mass appraisal is essentially the valuation of a group of properties with similar characteristics, which may be distinguished from the valuation of properties individually. Mangioni (2006) refers to the New South Wales Valuer General's approach to mass appraisal which is based on componentisation, being the process of grouping like properties, predominantly based on their location, use and physical attributes, into components. Local Government areas are broken down into a series of components within each of which a benchmark property is identified that is representative of a typical property within that component.

For example, a large number of like residential properties within a suburb may be grouped as a component and a typical residential property within that group identified as the benchmark property. The land comprising the benchmark property would then be valued annually (say $500,000 at 1 July 2023 and $550,000 at 1 July 2024) and the percentage movement in value since the previous valuation calculated (say +10%) which is then applied to the previous valuation (at, say, 1 July 2023) of all residential properties in that component to determine the updated value (at, say, 1 July 2024). To ensure consistency, a range of statistical checks are applied and any anomalies investigated. In the event of idiosyncratic differences in particular properties, these valuations may be 'handcrafted' to reflect such differences (Mangioni, 2006).

The mass appraisal process in New South Wales is explained graphically in Valuer General (2024), as shown in Figures 8.1 and 8.2.

While there may be issues in mass appraisal valuation approaches for statutory valuation, generally arising from mass appraisal approaches being a balance between cost and accuracy,

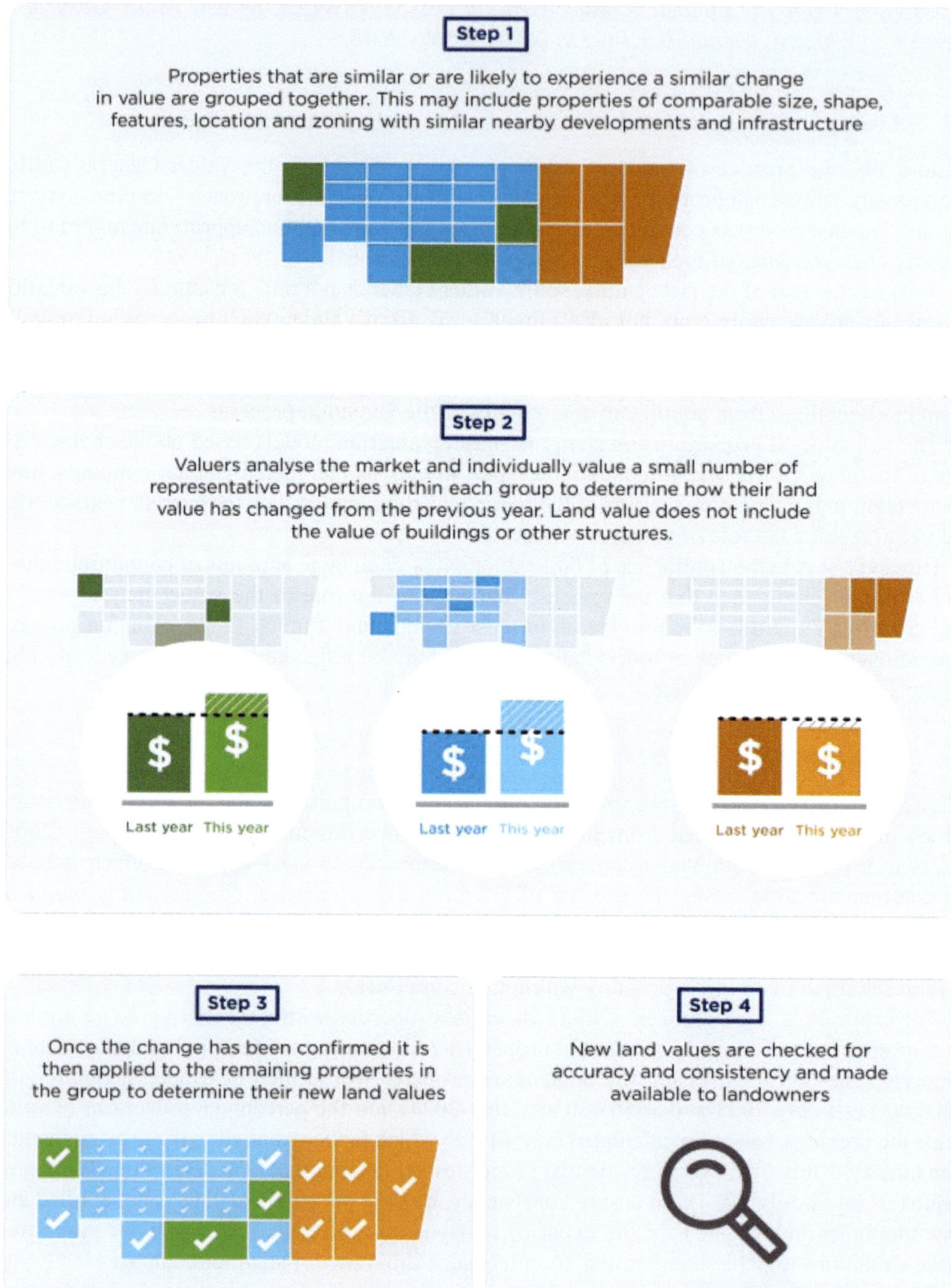

Figure 8.1 Mass appraisal valuation process in New South Wales.

Source: Valuer General (2024)

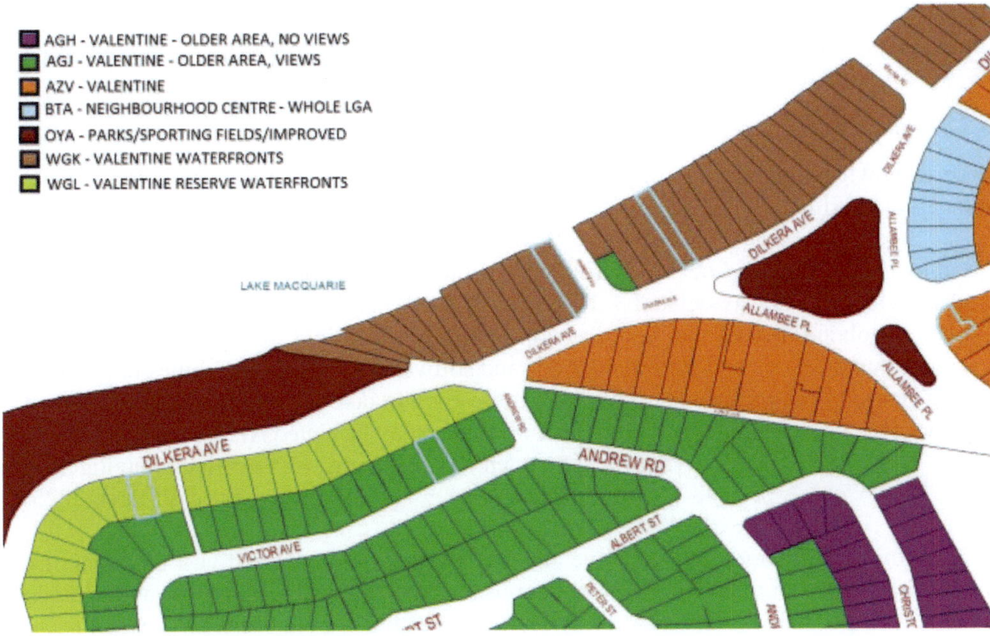

Figure 8.2 Componentisation within mass appraisal valuation process in New South Wales.

Source: Valuer General (2024)

over time such issues should be progressively resolved achieving community understanding of and confidence in the mass appraisal valuation system.

8.3.2 Automated valuation models

As Parker (2024) notes, automated valuation models seek to relate data representing the attributes of comparable properties to their sale prices in order to forecast the likely sale price of a subject property, often using hedonic modelling such as multiple regression analysis. Such modelling is most easily understood in the context of residential property, where the number of bedrooms, bathrooms, garages and so on, may be used to model the estimated sale price.

Automated valuation models are most effective for homogenous markets and so suit, for example, broad acre residential subdivisions or rows of townhouses, but are less suitable for heterogenous markets such as residential property with water views or commercial, retail or industrial property.

While automated valuation models based on multiple regression analysis have been taught in the valuation courses of the world's leading universities for over 40 years, it is only in the last two decades that the availability and quality of residential property data has increased sufficiently to make such models effective. Within Australia, there are several researchers, software houses and major lenders developing and using automated valuation models to provide assessments of value upon which residential mortgage lending credit decisions may be based.

While the use of automated valuation models for the assessment of capital value of improved residential property may be likely to be widespread within the foreseeable future, the use for statutory valuation is more challenging if the basis of value is land value. The limited number of land sales require the analysis of house sales to derive land value which currently still requires the judgmental input of valuers, as does the assessment of land value for non-residential property.

8.3.3 Artificial intelligence

While acknowledging the credibility limitations of Wikipedia, the following provides a succinct explanation of artificial intelligence:

> *Artificial intelligence (AI)* refers to the capability of computational systems to perform tasks typically associated with human intelligence, such as learning, reasoning, problem-solving, perception, and decision-making. It is a field of research in computer science that develops and studies methods and software that enable machines to perceive their environment and use learning and intelligence to take actions that maximize their chances of achieving defined goals. Such machines may be called AIs.
>
> (Source: https://en.wikipedia.org/wiki/Artificial_intelligence, accessed 6 March 2025)

While automated valuation models have already successfully converted vast amounts of data into assessments of capital value for homogenous residential property, it may be anticipated that artificial intelligence will eventually facilitate the conversion of even vaster amounts of data into assessments of capital value, rental value and land value for the principal property sectors for statutory valuation purposes.

As Parker (2024) notes, the emergence of artificial intelligence provides the Australian statutory valuation profession with an exciting opportunity to plan for its future by identifying those valuation services most at risk and diversifying their business model to embrace other valuation services less at risk. Central to the successful application of artificial intelligence to property valuation is the effective replication of the roles of judgment and local knowledge, both of which may be expected to be very difficult to achieve.

The roles of judgment and local knowledge may be considered through the example of a residential property with water views where large quantities of high-quality data in an artificial intelligence application may be unlikely to usurp the role of valuer judgment on the impact of view on value and the valuer's local knowledge of the comparable properties' relative views, aspect and accessibility.

8.4 Impact of major disasters

Across Australia and globally, major disasters are occurring more regularly and awareness of them and their impact is much greater through social media. Such major disasters may be local and natural, such as floods or bushfires, or may be global and transmissible, such as the COVID-19 pandemic.

8.4.1 Major disasters – principles

For statutory valuation following a major disaster, the principal issues are:

- timing of the major disaster;
- extent of the major disaster and
- availability of data.

8.4.1.1 *Timing*

Statutory valuation generally follows a pre-determined cycle of valuation dates, for example an annual valuation date of 1 July. The operation of the valuation cycle requires that there be sales evidence in the market for the period leading to 1 July, upon which valuations can be based. Further, the operation of the valuation cycle requires that there be comparable evidence for the subject property being valued.

If a major disaster occurs within a few months prior to the valuation date, there may be no comparable sales evidence for the subject property being valued as at the valuation date. For example, if a major disaster occurs on 1 April, such as a bushfire, there may be no comparable sales evidence of bushfire affected blocks as at the valuation date of 1 July.

However, the community through social media will not accept that the owner of a bushfire affected block receives a notice of valuation as at 1 July showing a valuation of a block unaffected by bushfire. While the Valuer General may have no comparable sales evidence of bushfire affected blocks upon which to base a valuation, the community through social media will not accept that as a valid reason for not reflecting the bushfire impact (usually a reduction) on the valuation of the impacted block.

Accordingly, the community through social media expectations are that the Valuer General will take steps to ensure fairness and equity in the valuation provided to the bushfire impacted block owner. However, while the relevant statutes may include provision for the valuation of individual blocks by the Valuer General and Court precedent permits the use of alternative valuation methods in the absence of comparable sales, community through social media expectations are for a pro-active and prompt response by the Valuer General to the impact of a major disaster.

8.4.1.2 *Extent*

Conventionally, Valuers General will divide a State or Territory into a series of valuation areas, either for valuation using mass appraisal methods or for valuation by teams of valuers. A natural disaster may not confine itself conveniently within one of the designated valuation areas and the community through social media will not accept that fairness and equity be denied to a landowner impacted by the major disaster simply because their block is inconveniently located. Accordingly, the Valuer General may have to adapt the approach to valuation to accommodate the extent of a major disaster.

8.4.1.3 *Availability of data*

It is common following local and natural major disasters, such as floods or bushfires, for there to be no sales of affected property for several months as landowners assess the situation, talk with their insurers and decide what to do. Following a global and transmissible major disaster, such as the COVID-19 pandemic, there were no sales in some sectors of the property market whereas sales continued in other sectors.

Despite the absence of sales, following the *Spencer* definition, the Valuer General is still required to assess what a hypothetical willing buyer would pay a hypothetical willing seller for the impacted block. In many cases, the Valuer General will be assessing land value, either as required under the relevant statute or because the house and/or improvements have been destroyed. While it may be contended that land value is not impacted by the major disaster, if it is a local and natural disaster there will be likely to be a value effect from stigma and absence of potential buyers and if it is a global transmissible disaster there may be a value effect from an absence of potential buyers.

8.4.2 Major disasters – practice

The Valuer General of New South Wales considered the impact of the following major disasters on land value:

- bushfire;
- flooding and
- COVID-19.

8.4.2.1 Bushfire

In Valuer General (2020a), the Valuer General of New South Wales analysed the impact of specific previous bushfires on land values in New South Wales in order to form a view on the potential impact of the 2019/2020 bushfires on land values as at 1 July 2020.

Four historic bushfire impacted areas were selected based on the degree of damage from the fire, availability of pre- and post-fire sales evidence and to cover a mix of land uses including urban residential, rural residential and rural. The four impacted areas selected were the Blue Mountains bushfire in 2013, the Carwoola bushfire in 2017, the Sir Ivan bushfire in 2017 and the Tathra bushfire in 2018.

A median price analysis and an analysis of individual sales before and after the bushfire was undertaken for each of the four selected impacted areas. Significantly, the review did not find consistent changes across the four areas analysed with no consistent land value reduction found in Tathra and the following changes observed in other areas:

- –2% to –6% in the Blue Mountains;
- –3% to –8% in Carwoola and
- –11% to –33% in Sir Ivan.

The review identified that three variables required consideration when determining the impact of bushfires on land values, being:

- the number of properties destroyed as a proportion of the total number of properties in the locality;
- the prevailing level of demand for vacant land in the locality before the bushfires, particularly if the bushfires result in an oversupply of land leading to a reduction in values and
- the specific location of the subject property within the locality, with properties located outside the immediate fire affected area expected to be less impacted.

To provide fairness and equity in the 2020 land values for bushfire impacted blocks, in the absence of comparable sales evidence the Valuer General took the unusual and innovative step of applying a percentage discount to the 2019 land values.

Having regard to the three variables, the percentage discounts applied to bushfire impacted blocks by the Valuer General are summarised in Table 8.1.

Accordingly, the Valuer General addressed the principal issue of timing and of unavailability of data by applying a percentage discount to the 2019 land values and of extent by considering properties located within the immediate fire affected area having regard to the extent of the fire, topography and location of specific properties.

Table 8.1 Percentage discounts in absence of comparable sales

Proportion of properties destroyed in area	*Prevailing vacant land demand*	*Change to land value*
High (over 50%)	Low	−30%
	Medium	−20%
	High	−10%
Medium (25%–50%)	Low	−25%
	Medium	−15%
	High	−5%
Low (under 25%)	Low	−20%
	Medium	−10%
	High	0%

Source: Valuer General (2020a)

8.4.2.2 Flooding

In Valuer General (2022), the Valuer General of New South Wales analysed the impact of severe weather and flooding across New South Wales between February and July 2022, focusing on two heavily impacted areas being Northern New South Wales (Lismore, Richmond Valley, Ballina, Kyogle, Byron, Clarence Valley and Tweed LGAs) and the Hawkesbury LGA.

For Northern New South Wales, which was impacted by flooding in February and March 2022, Valuer General (2022) found that sufficient sales occurred with which to undertake the 1 July 2022 valuation, with post-flood sales evidence showing the following relative to 1 July 2021 valuations:

- in areas with limited flood impact there was no impact on land values with some areas increasing in value;
- reductions of up to 10% in moderately impacted areas and
- reductions of up to 35% in the more significantly impacted areas, predominately in the Lismore area.

For the Hawkesbury LGA, which was impacted by flooding in July 2022, Valuer General (2022) found sufficient sales occurred with which to undertake the 1 July 2022 valuation except for an area along the Hawkesbury River, between Richmond and Wilberforce, where properties were fully inundated and there was no sales evidence in the area to demonstrate the impact on land values. Having regard to findings in Northern New South Wales and discussions with local agents in September 2022, Valuer General (2022) applied a 20% reduction to the 1 July 2021 land values for such properties, consistent with the approach adopted in Valuer General (2020a).

Accordingly, the Valuer General addressed the principal issue of timing and of unavailability of data by applying a percentage discount to the 2021 land values and of extent by only considering properties that were fully inundated and for which there was no sales evidence.

8.4.2.3 COVID-19

In Valuer General (2020b), the Valuer General of New South Wales analysed the impact of COVID-19 restrictions on the New South Wales property market, focusing on the period 1

March 2020 to 31 May 2020 which included the period when on-site auctions and inspection of property were banned.

As Valuer General (2020b) noted, the impact of COVID-19 on the New South Wales property market was profound:

> The impact of COVID-19 has caused economic consequences to business operations and finance, including changes to the labour workforce environment, travel restrictions within Australia and international tourism to Australia. These disruptions caused a re-evaluation by participants in the property market as they consider the purpose and value of real estate in business, as a place to live, or as investment holdings.

With the valuation of all landholdings in New South Wales due on 1 July 2020, the analysis in Valuer General (2020b) sought to determine:

- if the property market in NSW was sufficiently active at 1 July 2020 to support the 2020 valuation;
- if the residential and non-residential property market sectors of the NSW property market had been impacted to a similar extent by COVID-19; and
- if there were any implications for statutory valuation practice arising out of COVID-19.

The Valuer General determined that the NSW property market was sufficiently active overall to support a 1 July 2020 valuation, having analysed the level of transactions in:

- the pre-COVID-19 period (July 2019–February 2020) and
- in the COVID-19 period (March 2020–May 2020), when restrictions were imposed; and
- compared the findings to the July 2018–May 2019 period,

and, in an unusual and innovative step, having held discussions with major owners and representative industry bodies active in the property market.

However, the Valuer General determined that, while the residential property market in NSW remained active during COVID-19, transaction levels in the non-residential property market (including office, retail, industrial, hotels and so forth) were significantly lower in the COVID-19 period than in same period in the previous year. Therefore, all sectors of the NSW property market had not been impacted to a similar extent by COVID-19.

While there were very few transactions in the non-residential property market, COVID-19 may have been expected to adversely impact revenue and capital value for non-residential property and so to impact land value. To gauge the possible level of adverse impact, the Valuer General took the unusual and innovative step of holding discussions with major owners and representative industry bodies to obtain views on the effect of COVID-19 on revenue, capital value and land value for improved properties and development sites.

These discussions identified that different sectors of the non-residential property market had been impacted to differing degrees, which were then classified as high, medium and low impact. For high and medium impact, given the absence of comparable sales, the Valuer General applied a hypothetical development approach on a before and after basis to estimate the potential percentage fall in land value between 1 July 2019 and 1 July 2020, consistent with the approach adopted for bushfire and flood impacted landholdings, as summarised in Table 8.2.

Table 8.2 Valuer General percentage reductions in land value between 1 July 2019 and 1 July 2020 (conditions apply)

Sub-sector	*% land value reduction*
Industrial	0%
Regional motel	0%
Metro and regional licensed premises	0%
Caravan parks	0%
Childcare	0%
Rural	0%
Other regional office	0%
Larger regional office (Newcastle, Wollongong)	–10%
Neighbourhood shopping centre	–10%
Freestanding car park	–10%
Regional-based quarries	–10%
Prime CBD office	–15%
Secondary CBD office	–15%
Sub-regional shopping centre	–15%
Self-storage	–15%
Coal mines	–15%
Regional shopping centre	–20%
CBD and metro hotels	–20%
Metro-based quarries	–20%
Serviced apartment	–25%

Source: Valuer General (2020b)

The adoption of the hypothetical development method as the principal method of valuation in the absence of comparable sales was identified by the Valuer General as an implication for statutory valuation practice arising out of COVID-19.

Accordingly, for the valuation as at 1 July 2020, the Valuer General applied the percentage deductions shown in Table 8.2 to the land values determined at 1 July 2019.

Therefore, the Valuer General addressed the principal issue of timing and of unavailability of data by applying an alternative valuation method to determine a percentage discount to the 2019 land values and of extent by consulting directly with market participants in the non-residential property market for which there was no sales evidence.

8.4.3 Summary – impact of major disasters

As major disasters occur more frequently and more variedly, with the increasing role of social media reshaping the community tolerance for delay in response from Government, Valuer Generals need to be more innovative and embrace approaches not considered in the past, provided such approaches are not prohibited by the relevant statute and relevant Court decisions.

The New South Wales Valuer General's adoption of direct discussions with property market participants, valuation methods other than reliance on comparable sales and the adoption of evidence based percentage discounts to the previous year's valuation provide examples of the innovative statutory valuation responses that can be adopted to determine land values in the event of natural disasters.

8.5 Litigation and expert witness

Section 7.4.8 in Chapter 7 noted that, generally, the relevant statutes will provide for an appeal process if the landowner is dissatisfied with the outcome of an objection to a land value assessment with an appeal to be lodged within a defined timeframe. For example, in Victoria the right of appeal is to the Victorian Civil and Administrative Tribunal (VCAT), in Queensland to the Land Court of Queensland and in New South Wales to the Land & Environment Court.

Both the landowner and the Valuer General will usually retain a valuer to act as an expert witness in Court proceedings, with the role of an expert witness and the litigation process considered further below.

8.5.1 Role of the expert witness

The role of the valuer acting as an expert witness is to provide professional advice to the Court that is independent of the instructing party, with the duty of the valuer being to the Court. Therefore, the valuer as an expert witness is not acting on behalf of the instructing party, though they will be paid by the instructing party, but is acting on behalf of the Court.

The valuer as expert witness is not, therefore, an advocate for the instructing party. The valuer as expert witness has to make up their own mind on the value of the landholding or property in dispute and not be guided by or influenced by the instructing party.

The *API Rules of Professional Conduct* (API, 2024) require the strict maintenance of independence and impartiality when providing professional services stating that a member cannot act as an advocate and an expert in the same matter or as an advocate/expert when another member of the same firm is acting as an advocate/expert in the same matter.

Judicial consideration of what constitutes an expert witness may be found in *Makita (Australia) Pty Ltd v Sprowles* [2001] NSWCA 305. Essentially, an expert is a person who has specialised knowledge based on that person's training, study or experience with the primary duty of experts in giving opinion evidence being to furnish the trier of fact with criteria enabling evaluation of the validity of the expert's conclusion, such as comparable sales supporting an opinion of value. Further, the opinion expressed by the expert witness must be wholly or substantially based on that specialised knowledge (*Dasreef Pty Ltd v Hawchar* [2011] HCA 21).

In the case of a valuer acting as an expert witness, specialised knowledge will generally comprise a recognised property or valuation degree, professional body membership, State registration (if required) and a demonstrable period of experience valuing the type of property that is the subject of the litigation.

Most Courts will require a valuer's CV to be attached to their valuation report, with the judicial officer usually reviewing the CV for evidence of specialised knowledge based on training, study or experience. As the effectiveness of the valuer as an expert witness is dependent on credibility, it is essential that the CV attached to the valuation report supports the claim of expertise. For example, the credibility of a valuer who usually values houses in the western suburbs of Sydney will be in question if attempting to act as an expert witness in the valuation of shopping centres in the eastern suburbs. It is, therefore, important for the valuer to only provide expert advice in areas where they have direct, relevant experience.

8.5.1.1 Expert witness – duties

The API guidance paper, *ANZPGP 203 Acting as an expert witness* (API, 2021), states the duties of an expert witness to be as follows:

> *1.5 Duties of an Expert witness*
>
> An expert witness has several conflicting duties which need to be managed and resolved correctly. The most important is the duty to the Court and is to be treated as a paramount and overriding duty. At its core is an obligation to assist the Court in arriving at the truth. It should be borne in mind that, it is the Court, which makes the ultimate decision, and the role of the expert is to educate the Court to the same level of understanding as the expert on the issue in question. The expert is not an advocate for a particular party and should remain true to their profession and refrain from attempting 'to win' the case for the client. However, an expert witness is to ensure that all of the relevant points of their client's case are brought to the attention of the Court.
>
> The expert has a duty to maintain their reputation by providing competent advice to the Court (reasoned conclusions are preferred to bold conclusions) and refusing to put forward arguments that lack substance or credibility to unreasonably advance the client's case. It is inappropriate to allow the instructing lawyer or any outside party to 'filter' the expert's report. The lawyer may provide advice on legal or procedural matters and may even provide advice as to the scope of the report and its relevance to the proceedings. With Joint Conferencing it is inappropriate to allow the instructing lawyer or any outside party to interfere with the independent expert witness process once a Joint Conference has commenced. The exception to this is where a facilitator is engaged to assist the experts to produce a credible Joint Report.
>
> The duty to the client includes confidentiality, diligence especially in presenting the case fully and fairly, punctuality, not to mislead on the prospects of success to obtain instructions and to quote fairly and transparently.
>
> The duty is to be truthful as to fact, honest as to opinion and complete as to coverage of relevant matters. The duty is the same whether the Expert is giving evidence in Court or to a Tribunal on oath or not on oath. Expert evidence must be independent, objective, and unbiased.

Further, API (2021) notes:

> *1.6 Code of Conduct*
>
> The various Practice Notes published by the Courts and legislation often refer to a 'code of conduct' which usually sets out specifically, the expert's duties, reporting requirements and includes an undertaking for the expert to have read and agree to be bound by the code.

considered further in Section 8.5.2.

8.5.1.2 Expert witness – instructions

The solicitor for the applicant or respondent will usually contact the valuer to ascertain if they may be able and willing to act in the matter. Prior to accepting instruction, the valuer must ensure that they have the specialised knowledge to provide the expert opinion required and have no conflicts of interest. If so, the solicitor will usually provide the valuer with formal instructions detailing the expert opinion required, often framed as a series of questions for response, together with a brief of relevant documents. Further information on instructions may be found in API (2021), section 1.8.

8.5.1.3 Expert witness – statement of evidence

The valuer will then read the brief of relevant documents thoroughly and prepare an expert opinion responding to the questions raised in the formal instructions. As the expert report will be thoroughly analysed by the other party in the litigation, it is important that it is fully supported by factual evidence, such as comparable sales evidence rather than statements of opinion or hearsay evidence, within the report. Accordingly, it is common for expert reports to be over a hundred pages long with a large number of appendices. Further information on report preparation may be found in API (2021), Sections 2.0, 3.0 and 4.0.

Some States and Territories may require the expert valuer witnesses for both sides to confer in an effort to narrow the areas of differences and reach agreement, usually resulting in the preparation of a joint expert report recording the agreement and disagreement achieved. Further information on joint conferencing, joint reports and joint evidence may be found in API (2021), section 7.0, 8.0 and 9.0.

8.5.1.4 Expert evidence – presentation and delivery

At the hearing in Court, the valuer will attend to present their report and respond to questions from the barristers for both sides. On the day of the hearing, prior to giving evidence, the valuer will probably have to sit and wait in Court as the matter unfolds. Court etiquette usually requires that:

- mobile phones are turned off;
- there is no eating and drinking;
- witnesses sit quietly and do not interrupt proceedings;
- no photographs are taken;
- no audio or video recordings are taken; and
- care is taken in any conversation with witnesses for the other party.

When presenting and delivering expert evidence, it is essential that the valuer knows their report thoroughly and is able to clearly and fully articulate the contents of the report and the underlying rationale. To be perceived as a credible witness by the Court, when giving evidence the valuer should:

- review the question asked by a solicitor or barrister in their mind, making sure that they fully understand it;
- never answer a question that they do not fully understand, asking for clarification if necessary;
- consider their answer carefully before giving it, not being afraid to take two to three minutes before providing an answer;

- refer to notes and documents where necessary;
- talk to the judge and not to the lawyers;
- keep answers short and to the point in cross examination;
- never guess answers, stating if they do not know the answer or cannot remember it;
- never lose their temper or become emotionally involved, attempting to avoid emotional responses of any nature;
- always pause to give their lawyer opportunity to object;
- watch for double questions;
- if caught with a mistake, admit it and correct it, not trying to defend their mistake;
- be positive and assertive in their approach, avoiding tentative answers such as 'I think I used' or 'possibly I considered';
- testify honestly and sincerely and
- dress and present themselves appropriately.

Further information on presentation may be found in API (2021), section 1.10, 5.0 and 6.0.

Acting as an expert valuer in Court proceedings can be both intellectually challenging and stressful, with not everybody being suited to the role. As with many forms of specialist valuations, some valuers develop a practice as an expert witness with repeated instructions allowing them to refine their skills in expert report preparation, provision of expert evidence in Court and managing examination by barristers.

8.5.2 Litigation process

Some State and Territory Courts provide practice notes to guide expert witnesses concerning the preparation of their expert reports which are considered below, followed by a brief overview of the Court process.

8.5.2.1 Practice notes

Those State and Territory Courts that publish practice notes may also refer to a code of conduct or similar document. For example, the New South Wales Land and Environment Court adopts the expert witness code of conduct provided in the *Uniform Civil Procedure Rules* 2005 which state:

> Schedule 7 Expert witness code of conduct
>
> 1 *Application of code*
> This code of conduct applies to any expert witness engaged or appointed –
>
> (a) to provide an expert's report for use as evidence in proceedings or proposed proceedings, or
> (b) to give opinion evidence in proceedings or proposed proceedings.
>
> 2 *General duties to the Court*
> An expert witness is not an advocate for a party and has a paramount duty, overriding any duty to the party to the proceedings or other person retaining the expert witness, to assist the court impartially on matters relevant to the area of expertise of the witness.

3 *Content of report*

Every report prepared by an expert witness for use in court must clearly state the opinion or opinions of the expert and must state, specify or provide –

(a) the name and address of the expert, and
(b) an acknowledgement that the expert has read this code and agrees to be bound by it, and
(c) the qualifications of the expert to prepare the report, and
(d) the assumptions and material facts on which each opinion expressed in the report is based (a letter of instructions may be annexed), and
(e) the reasons for and any literature or other materials utilised in support of each such opinion, and
(f) (if applicable) that a particular question, issue or matter falls outside the expert's field of expertise, and
(g) any examinations, tests or other investigations on which the expert has relied, identifying the person who carried them out and that person's qualifications, and
(h) the extent to which any opinion which the expert has expressed involves the acceptance of another person's opinion, the identification of that other person and the opinion expressed by that other person, and
(i) a declaration that the expert has made all the inquiries which the expert believes are desirable and appropriate (save for any matters identified explicitly in the report), and that no matters of significance which the expert regards as relevant have, to the knowledge of the expert, been withheld from the court, and
(j) any qualification of an opinion expressed in the report without which the report is or may be incomplete or inaccurate, and
(k) whether any opinion expressed in the report is not a concluded opinion because of insufficient research or insufficient data or for any other reason, and
(l) where the report is lengthy or complex, a brief summary of the report at the beginning of the report.

4 *Supplementary report following change of opinion*

(1) Where an expert witness has provided to a party (or that party's legal representative) a report for use in court, and the expert thereafter changes his or her opinion on a material matter, the expert must forthwith provide to the party (or that party's legal representative) a supplementary report which must state, specify or provide the information referred to in clause 3(a), (d), (e), (g), (h), (i), (j), (k) and (l), and if applicable, clause 3(f).
(2) In any subsequent report (whether prepared in accordance with subclause (1) or not), the expert may refer to material contained in the earlier report without repeating it.

5 *Duty to comply with the court's directions*

If directed to do so by the court, an expert witness must –

(a) confer with any other expert witness, and
(b) provide the court with a joint report specifying (as the case requires) matters agreed and matters not agreed and the reasons for the experts not agreeing, and
(c) abide in a timely way by any direction of the court.

6 *Conferences of experts*
Each expert witness must –

(a) exercise his or her independent judgment in relation to every conference in which the expert participates pursuant to a direction of the court and in relation to each report thereafter provided, and must not act on any instruction or request to withhold or avoid agreement, and
(b) endeavour to reach agreement with the other expert witness (or witnesses) on any issue in dispute between them, or failing agreement, endeavour to identify and clarify the basis of disagreement on the issues which are in dispute.

It is essential that the expert witness valuer follow carefully and adhere closely to the practice notes and code of conduct of the Court in which they are giving evidence.

8.5.2.2 Court process

Following instructions, each expert valuer will prepare their report and provide it to their instructing solicitor. Reports by the expert valuer for each party to the litigation will then be exchanged and each valuer will read and critique the expert witness report of the other valuer.

The expert valuers may then meet and confer, endeavouring to reach agreement on differences in opinion and identifying differences upon which they are unable to reach agreement. They will then jointly settle a joint expert's report summarising the issues in agreement and disagreement.

Prior to a hearing in Court, some States and Territories provide for an alternative dispute resolution procedure whereby the parties come together and through mediation or conciliation attempt to reach agreement in some or all of the matters in dispute. Each expert valuer will usually attend the conference and contribute to the attempted resolution of valuation issues. If the parties are unable to reach agreement, the matter will be listed for a hearing at a later date.

A hearing may be before a Judge and/or Commissioner or Member, depending on the State or Territory, generally proceeding as follows:

- the barrister or solicitor for the applicant (usually the ratepayer or land tax payer) will make an opening submission;
- the barrister or solicitor for the respondent (usually the Valuer General) will make an opening submission;
- the barrister or solicitor for the applicant will call expert witnesses and ask them questions about their expert report (examination);
- the barrister or solicitor for the respondent will ask each expert witness questions about their expert report and their previous answers (cross examination);
- the barrister or solicitor for the applicant will ask each expert witness about matters arising from their answers to the respondent's barrister or solicitor;
- the above three steps will be repeated by the barrister or solicitor for the respondent and applicant with each of the respondent's witnesses called;
- the barrister or solicitor for the applicant will make a closing submission, summing up their client's contentions;
- the barrister or solicitor for the respondent will make a closing submission, summing up their client's contentions;

- the barrister or solicitor for the applicant will make a submission on matters that may have been raised in the respondent's closing submission;
- the Judge and/or Commissioner or Member may issue a judgment immediately (ex tempore) or reserve judgment (reserved), close the hearing and issue a judgment at some later point.

When in the witness box, the valuer as an expert witness should strive to appear credible, being of smart and professional appearance, speaking clearly and assertively and presenting a demeanour of confident humility. Depending on the issues in dispute and the complexity of the matter, a hearing may be completed in a day or require several days for completion.

8.6 Summary and conclusions

Focusing on the key issues from a valuer's perspective, Chapters 1 and 2 introduced the concepts and principles of statutory valuation, with statutory valuation for acquisition considered in Chapters 3, 4, 5 and 6. Statutory valuation for rating and taxing was considered in Chapters 7 and 8, with regulatory valuation then considered in Chapters 9 and 10.

Chapter 1 introduced the principles of statutory valuation, including the concepts of market value, liberal/conservative estimates of value, highest and best use, potentiality and confirmation of a foresight. This provided the essential foundation principles of statutory valuation that are then applied in the following chapters on statutory valuation.

Chapter 2 considered the practice of statutory valuation including valuation methods with an emphasis on the comparative method of valuation, together with statutory valuation conventions such as the before and after method and the piecemeal method. This provided an understanding of the valuation methods and conventions that may be preferred for application in the following chapters on statutory valuation.

Statutory valuation for acquisition is considered in Chapters 3, 4, 5 and 6. Chapter 3 introduced the principles of statutory valuation for acquisition, including the statutory framework and process, with a focus on the concept of disregard of the scheme in compensation assessment and the distinction between concepts of value and worth. This provided an insight into concepts of fairness and equity in statutory valuation in the context of the disregard of compulsion.

Chapter 4 explored the market value bases of compensation in acquisition for which statutory valuation may be required, including market value of the interest and the related market value compensation concepts of severance and injurious affection. This provided a broad understanding of the heads of compensation for acquisition on a value basis that may arise in statutory valuation.

Chapter 5 considered non-market value forms of compensation including special value which was noted to rarely arise, payment for solace arising from loss of home which is heavily dependent on the exact wording of the facilitating statute and the concept of disturbance for costs and losses suffered as a result of the acquisition. This provided a broad understanding of the heads of compensation for acquisition that are not, necessarily, on a value basis but that may arise in statutory valuation.

Chapter 6 concluded the chapters considering statutory valuation for acquisition by examining compensation for other interests such as easements and mortgages, mineral rights and profits à prendre, strata title and substrata, businesses and the emerging areas of native title and of indigenous cultural loss. This provided an appreciation of the diversity of interests that may be subject to acquisition and so may require valuation for the assessment of compensation.

Statutory valuation for rating and taxing was considered in Chapters 7 and 8. Chapter 7 introduced the principles of land as the basis for taxation through land tax, stamp duty and rates, considering the general principles of the valuation of land for taxation using the New South Wales statutory framework as an example. This provided a general understanding of the often arcane and artificial basis of statutory valuation upon which Government collects very significant amounts of taxation revenue.

This chapter explored the practice of statutory valuation for rating and taxing through exemptions, concessions and allowances that have evolved over the last century. Such evolution has followed both changes in society, such as demographic changes leading to a growing elderly population, together with political changes in response to the cost and availability of housing. The combined effect of the evolution of exemptions, concessions and allowances is to progressively complicate that which started as a relatively simple taxation system with the risk of consequential distortions to the underlying property market.

Contemporary issues such as mass appraisal, automated valuation models and artificial intelligence were considered with the confluence of data availability and computing power potentially threatening the role of the valuer, unless the valuer adapts to focus on the idiosyncratic features of individual properties that may be beyond the scope of computer based valuation to address.

Following an analysis of how statutory valuation may respond to major disasters in the age of intolerant social media, the role of the valuer in litigation and the importance of professional credibility as an expert witness were considered. This chapter provided an insight into the continued evolution of statutory valuation to address issues as they arise over time, highlighting the importance of a principles-based approach to statutory valuation.

Regulatory valuation, considered in the context of the preparation of valuations for inclusion in financial reporting, is considered in Chapters 9 and 10. Chapter 9 explores the principles of regulatory valuation with regard to the inter-related roles of international accounting standards, international financial reporting standards and international valuation standards, focusing on the primacy of fair value in international financial reporting standards. This provides an understanding of the fundamental conceptual differences and bases between statutory valuation and regulatory valuation.

Finally, Chapter 10 concludes regulatory valuation by exploring the practice of valuation approaches and methods relevant for regulatory valuation such as the income approach and the cost approach and the various valuation methods there within. This provides an appreciation of the wider range of valuation approaches and methods available to the valuer for regulatory valuation reflecting that, unlike statutory valuation, there is no requirement for defence in Court (negligence claims excepted).

References

API (2021) *ANZPGP 203 Acting as an expert witness*, Australian Property Institute, Deakin

API (2024) *Rules of professional conduct*, Australian Property Institute, Deakin

Mangioni, V (2006) *Land tax in Australia*, Australian Property Publications, Bondi

Parker, D (Ed) (2022) *Principles and practice of property valuation in Australia*, Routledge, Abingdon

Parker, D (2024) *Introduction to property valuation in Australia*, Routledge, Abingdon

Valuer General (2020a) *Review of the impact of bushfires on land values*, Valuer General of New South Wales, Sydney

Valuer General (2020b) *Review of the impact of COVID-19 on the NSW property market*, Valuer General of New South Wales, Sydney

Valuer General (2022) *Review of the impact of flooding on the 1 July 2022 land values*, Valuer General of New South Wales, Sydney

Valuer General (2024) *Report on NSW land values at 1 July 2024*, Valuer General, Sydney

9 Regulatory valuation

Principles

This chapter draws heavily on the following, which are duly acknowledged here rather than individually referenced through the chapter, and to which readers are referred for a deeper consideration:

Parker, D (2016) *International valuation standards – a guide to the valuation of real property assets*, Routledge, Abingdon

Parker, D (2022) Valuation principles, in D Parker (Ed) *Principles and practice of property valuation in Australia*, Routledge, Abingdon

Parker, D (2024) *Introduction to property valuation in Australia*, Routledge, Abingdon

Stewart, R (2022) Valuation for financial reporting purposes, in D Parker (Ed) *Principles and practice of property valuation in Australia*, Routledge, Abingdon

9.1 Introduction

This book is an introduction to two very different valuation worlds in Australia – the complex and often arcane national realm of statutory valuation and the dynamic global empire of regulatory valuation. Statutory valuation is different from regulatory valuation, with the former being based on relevant applicable statutes with valuations being capable of challenge on appeal in Court while the latter is based on relevant applicable regulations with valuations being incapable of challenge in Court (negligence claims excepted).

As an introductory textbook on statutory and regulatory valuation, this book is a companion to the Australian valuation textbooks, *Introduction to Property Valuation in Australia* (Parker, 2024) and *Principles and Practice of Property Valuation in Australia* (Parker, 2022), which provide a wider analysis of key principles underlying property valuation and current techniques and issues in the practice of property valuation for major sectors of the Australian property market.

Statutory valuation in Australia is most commonly encountered in valuations for rating, taxing and acquisition but may arise wherever a statute prescribes a basis of or approach to valuation. Regulatory valuation is most commonly encountered in valuations for financial statements and taxation purposes but may arise wherever a regulation prescribes a basis of or approach to valuation.

Rating and taxing of land raise significant amounts of revenue for State Governments and Councils in Australia, based on statutory valuation. For example, the NSW 2025–26 Budget Statement estimates land tax revenue for 2024–2025 to be $8.2 billion with the City of Brisbane Annual Plan and Budget 2024–2025 estimating proposed rates and utility charges to be $1.51 billion.

DOI: 10.1201/9781032616957-9

Acquisition of land, property and businesses commonly arises for major infrastructure projects and may be by agreement or by compulsion, based on statutory valuation. For example, such major infrastructure projects in Australia currently include the multi-billion-dollar Sydney Metro, Melbourne Suburban Rail Loop and Melbourne to Brisbane Inland Rail projects.

Valuation of land and property holdings based on regulatory valuation arises for all reporting entities in Australia that publish financial statements upon which it is reasonable to expect that users will rely for decision making. For example, at June 2025 there are over 2,000 reporting entities listed on the Australian Stock Exchange which each publish financial statements, with the financial statements for Goodman Group, as at 30 June 2024, showing the fair value of total property assets to be $20.2 billion, based on regulatory valuation.

Focusing on the key issues from a valuer's perspective, the concepts and principles of statutory valuation are introduced in Chapters 1 and 2, with statutory valuation for acquisition considered in Chapters 3, 4, 5 and 6 and statutory valuation for rating and taxing considered in Chapters 7 and 8. Regulatory valuation is then considered in Chapters 9 and 10.

Chapter 1 introduced the principles of value in statutory valuation, including the concepts of market value, liberal/conservative estimates of value, highest and best use, potentiality and confirmation of a foresight. This provided the essential foundation principles of statutory valuation that are then applied in the following chapters on statutory valuation.

Chapter 2 considered the practice of statutory valuation including valuation methods for statutory valuation with an emphasis on the comparative method of valuation, together with statutory valuation conventions such as the before and after method and the piecemeal method. This provided an understanding of the valuation methods and conventions that may be preferred in the following chapters on statutory valuation.

Statutory valuation for land acquisition is considered in chapters 3, 4, 5 and 6, supported by an evolving example in Appendix 1 of each chapter. Chapter 3 introduced the principles of statutory valuation for land acquisition, including the statutory framework and process, with a focus on the concept of disregard of the scheme in compensation assessment and the distinction between concepts of value and worth. This provided an insight into concepts of fairness and equity in statutory valuation in the context of the requirement to disregard the compulsory nature of the acquisition.

Chapter 4 explored the market value bases of compensation in acquisition for which statutory valuation may be required, including market value of the interest and the related market value compensation concepts of severance and injurious affection. This provided a broad understanding of the heads of compensation for acquisition on a value basis that may arise in statutory valuation.

Chapter 5 considered non-market value forms of compensation including special value, which was noted to rarely arise, payment for solace arising from loss of the primary place of residence and the concept of disturbance for costs and losses suffered as a result of the acquisition. This provided a broad understanding of the heads of compensation for acquisition that are not necessarily on a value basis but which may arise in statutory valuation.

Chapter 6 concluded the chapters considering statutory valuation for acquisition by examining compensation for other interests such as easements and mortgages, mineral rights and profits à prendre, strata title and substrata, businesses and the emerging statutory valuation areas of native title and indigenous cultural loss. This provided an appreciation of the diversity of interests that may be subject to acquisition and so may require valuation for the assessment of compensation.

Statutory valuation for rating and taxing was considered in Chapters 7 and 8. Chapter 7 introduced the principles of land as the basis for taxation through land tax, stamp duty and rates, considering the general principles of the valuation of land for taxation using the New South Wales statutory framework as an example. This provided a general understanding of the often arcane and artificial basis of statutory valuation upon which government collects very significant amounts of taxation revenue.

Chapter 8 explored the practice of statutory valuation for rating and taxing through exemptions, concessions and allowances that have evolved over the last century, together with contemporary issues such as mass appraisal, automated valuation models and artificial intelligence, the impact of major disasters on valuation for rating and taxing and the litigation process including the valuer as an expert witness. This provided an insight into the continued evolution of statutory valuation to address issues as they arise over time, highlighting the importance of a principles-based approach to statutory valuation.

Regulatory valuation, considered in the context of the preparation of valuations for inclusion in financial reporting, is considered in Chapters 9 and 10. Regulatory valuation differs from statutory valuation in various ways including:

- an emphasis on accounting standards and valuation standards rather than statute and case law;
- an emphasis on fair value as defined in International Financial Reporting Standards rather than value as defined in *Spencer v Commonwealth* (1907) 5 CLR 418;
- an emphasis on valuation for financial reporting purposes rather than for compulsory acquisition or taxation purposes (considered in this chapter) and
- having a wide range of valuation approaches available for adoption rather than an emphasis on the market approach and the comparative method of valuation (considered in Chapter 10).

This chapter explores the principles of regulatory valuation with regard to the interrelated roles of international accounting standards, international financial reporting standards and international valuation standards, focusing on the primacy of fair value as defined in international financial reporting standards. This provides an understanding of the fundamental conceptual differences and bases between statutory valuation and regulatory valuation.

Finally, Chapter 10 concludes regulatory valuation by exploring the practice of valuation approaches and methods relevant for regulatory valuation such as the income approach and the cost approach and the various valuation methods there within. This provides an appreciation of the wider range of valuation approaches and methods available to the valuer for regulatory valuation reflecting that, unlike statutory valuation, there is no requirement for defence of the valuation in Court (negligence claims excepted).

This book is based on statute, case law and regulations as at March 2025 and those standards and guidance documents published in IVSC (2024) and RICS (2024) and on the API website (accessed January to March 2025). Given their nature, statutes, case law, regulations, standards and guidance documents are dynamic and change continuously. Accordingly, readers should not rely on this book as a current statement of a statute, case law position, regulation, standard or guidance document and should undertake their own enquiries to find the most recent version.

Accordingly, this chapter will now consider:

- standards;
- valuation professional body standards and guidance;
- fair value and
- valuation for financial reporting purposes.

9.2 Standards

Whereas statutory valuation is governed by statute and case law, regulatory valuation (being considered as the preparation of valuations for inclusion in financial reporting) is governed by accounting standards and valuation standards, with the principal standards for consideration being:

- International Financial Reporting Standards;
- International and Australian Accounting Standards;
- International Valuation Standards and
- professional body standards and guidance (considered in section 9.3), including:
  - RICS standards and
  - API guidance.

The emergence and development of standards was a direct response to globalisation in the second half of the last century. Following the Second World War, developments such as the Bretton Woods Agreement on international monetary policy, the emergence of container shipping and international air travel all facilitated global integration in trade and commerce. Such global integration was further facilitated by late twentieth-century developments in communications, computing and the advent of the internet, contributing to the currently very high level of interconnectedness between the world's major economies.

The growth of multinational corporations was further fostered by bilateral and later regional trade agreements between countries, with the formation of GATT and the World Trade Organisation making it economically feasible for global corporates such as Ford and Coca-Cola to operate across the world, or for European-based corporates to manufacture goods in China or India and sell to world markets in the USA or Asia.

Globalisation has reduced the significance of national and political borders and increased the significance of economic integration. The interconnectedness of the world's major economies was glaringly apparent during the Global Financial Crisis of mid-2007 to early 2009, when problems arising in the US property, banking and finance markets rapidly became problems for the property, banking and finance markets of each of the world's major economies, and again during COVID and the Russia – Ukraine War, when global supply chains were adversely impacted.

In the context of property, globalisation and the growth in multinational corporations has significantly increased the amount of property held by businesses in different countries for the purpose of business operations. Further, the evolution of sovereign wealth funds, pension funds, superannuation funds, real estate investment trusts (REITs) and other property fund managers investing internationally, outside their country of origin, has resulted in many of the office towers, shopping centres and logistics properties in major cities around the world now being owned by foreign entities.

To facilitate global commerce, standards provide a common basis for company accounting worldwide that is understandable, comparable, reliable and relevant for internal and external users across different countries – effectively enabling an *apples with apples* comparison of financial data for decision making.

For property, the application of common accounting and valuation standards internationally should result in the stated value of each property held by a multinational corporation anywhere in the world meaning the same thing, so facilitating the free flow of capital. For a relatively small country like Australia, this reduces a major barrier to investment by multinationals such as McDonalds or Philips through the provision of accounting transparency.

While purportedly harmonised, such that inconsistencies should be minimal, the provisions of International Financial Reporting Standards and International Accounting Standards take precedence over International Valuation Standards. Given their nature, standards are dynamic, being regularly updated, with the most recently published versions replacing previously published versions. Accordingly, readers should not rely on this chapter as a current statement of a standard and should visit www.ivsc.org, www.ifrs.org, www.rics.org or www.api.org.au to find the most recent and current version.

9.2.1 International Financial Reporting Standards (IFRSs)

International Financial Reporting Standards (IFRSs) set out recognition, measurement, presentation and disclosure requirements dealing with transactions and events that are important in general purpose financial statements (IFRS, 2025) prepared by an entity on a going concern basis.

The benefits of IFRSs, as a common basis for accounting worldwide, include the provision of high-quality, transparent and comparable information in financial reporting to help investors, other participants in the various capital markets of the world and other users of financial information make economic decisions (IFRS, 2025). This aids the flow of global capital between countries, supports national economies and improves international competitiveness, as well as reducing financial reporting costs, improving the quality of financial reporting and providing more useful information to decision makers.

From the viewpoint of property valuation, the key standard is IFRS 13 *Fair Value Measurement* (IFRS, 2025) with the key basis of value being fair value as defined therein rather than market value. IFRS 13 provides a framework for the measurement of fair value and the requirements for disclosure in financial statements, based on the fundamental premise that fair value is a market-based measurement, not an entity-specific measurement.

For the determination of fair value, the entity is required to determine all of the following:

- the identity of the asset;
- the highest and best use of the asset;
- market in which an orderly transaction would take place, being the principal or most advantageous market for the asset and
- the valuation techniques appropriate for measurement, having regard to the availability of data and the fair value hierarchy of inputs.

Highest and best use is determined from the viewpoint of market participants rather than that of the measuring entity, even if the entity does not intend to use the asset for its highest and best use. The nominated highest and best use must fulfil all the following criteria:

- physically possible – such as location or size, as may be taken into account by market participants when pricing the asset;
- legally permissible – such as any legal restrictions on the use of the asset, as may be taken into account by market participants when pricing the asset and
- financially feasible – such as generating adequate income or cash flows, taking into account the costs of conversion, to produce an investment return that may be required by market participants when pricing the asset.

The selection of an appropriate valuation technique is dependent on the availability of sufficient data and prioritising the use of observable inputs that are relevant over unobservable inputs.

IFRS 13 categorises inputs into three levels, being the fair value hierarchy of inputs in descending order of priority:

- level 1 – observable inputs, such as quoted prices (unadjusted) in active markets for identical assets, such as equity stocks traded on a stock market;
- level 2 – directly or indirectly observable inputs for similar assets in markets that are not active, such as interest rate swaps, options on exchange traded shares or licensing arrangements;
- level 3 – unobservable inputs where there is little, if any, market activity for the asset at the measurement date, such as heterogenous property traded on a property market.

Entities are required to disclose the level within which measurements are categorised and to provide quantitative information about significant unobservable inputs within level 3 in tabular or other format. Given the extent to which property falls within level 3, the disclosure requirements for entities around property valuations are potentially very extensive for assets reported at fair value.

The definition of fair value in IFRS 13 is considered further in Section 9.4.

9.2.2 International and Australian Accounting Standards

The International Accounting Standards Board produces a single set of financial reporting standards based on clearly articulated principles for adoption through convergence with national accounting standards, including Australian Accounting Standards.

Currently, around 100 countries have adopted IASs, including Australia, the European Union, Canada, Montenegro and Nepal. However, major economies such as the USA, China, Japan and India are yet to require adoption for listed companies.

Stewart (2022) succinctly notes:

> Valuers engaging in financial reporting work need to be familiar with . . . the key Accounting Standards that set the rules of the game for financial reporting valuations.
>
> (page 211)

In Australia, most valuations for companies and private sector entities will be governed by Australian Accounting Standards, with the public sector having its own standards, IPSAS, which are generally similar but with specific differences in relation to heritage and infrastructure assets (Stewart, 2022).

Table 9.1 identifies those Australian Accounting Standards for which valuation is a central concern (Stewart, 2022), with the most commonly arising for financial reporting purposes being AASB 116 *Property, Plant and Equipment* and AASB 140 *Investment Property* (which are considered in further detail in Sections 9.5.2 and 9.5.3).

Australian Accounting Standards are mandatory by law for reporting entities and so sit above international and national valuation standards, which are not mandatory by law but are binding upon valuers who are members of a relevant professional body.

Significantly, for regulatory valuation, valuations for financial reporting purposes are subject to review by auditors with Stewart (2022) noting that an auditor will need to be satisfied with:

- the valuer's competence, capability and objectivity in preparing the valuation report;
- how the valuer went about their work and
- whether the valuer's work contains appropriate evidence for the auditor to conclude that the item valued in their report is appropriately carried by the client at fair value.

Table 9.1 Australian Accounting Standards and valuation issues

AASB No.	*Title*	*Broad summary of valuation issue*
2	Share-based Payments	Securities issued as consideration for service to be fair valued (para 10, specially defined in Appendix B)
3	Business Combinations	All assets and liabilities to be fair valued and separately allocated as part of Purchase Price allocation on a business combination (para 18)
5	Non-current Assets Held for Sale and Discontinued Operations	Assets held for sale to be held at lower of carrying amount or fair value less costs to sell (para 15)
9	Financial Instruments	Financial instruments to be carried at fair value (para 5.2.1)
16	Leases (supersedes AASB 117 for periods on or after 1/1/2019)	Rights of use and leases can be fair valued at the present value of lease payments (para 26)
116	Property, Plant and Equipment	Assets can be revalued at fair value (para 31) and applies AASB 136 (para 63)
136	Impairment of Assets	Assets may be impaired, that is, written down if higher of fair value less costs to sell (para 25) and value in use (para 30) is less than carrying amount
138	Intangible Assets	Can revalue intangible assets at fair value, if an active market exists (para 75)
140	Investment Property	Investment property must be carried at fair value (para 33)
141	Agriculture	Biological assets (para 12, e.g. an orchard) and Agricultural produce (para 13, e.g. a crop) must be valued at fair value
1051	Land Under Roads	May elect to value land under roads at fair value (para 13)
1056	Superannuation Entities	Assets held by (and liabilities of) superannuation funds must be held at fair value (para 13)

Source: Stewart (2022)

In accounting, valuations may be used to establish the initial value of assets (though this is generally their cost), to revalue them periodically or to assess the value for impairment for the purposes of the preparation of the financial statements of an entity (Stewart, 2022). While the assessment of asset values is fundamental to the assessment of the financial position and performance of Australian organisations, which directly impacts not only their shareholders but also a large proportion of the Australian community whose superannuation funds invest in such organisations, financial statements are not held out to be statements of value.

9.2.3 International Valuation Standards

This chapter is based on *International Valuation Standards 2024* (IVSC, 2024) with reference to International Valuation Standards 2013 (IVSC, 2013) and readers should check www.ivsc.org for more recent versions of the various standards referred to herein. The role of International Valuation Standards (IVS) in valuation in Australia is considered at length in Parker (2024) and the following provides a brief overview based on Parker (2022).

While IVS are harmonised with IFRS and IAS, the latter have precedence over the former in the event of inconsistency. Similarly, the RICS Red Book and API guidance are harmonised with IVS, with IVS standards having replaced API standards in Australia.

As a heavily simplified summary, IVS may be considered to comprise three key concepts (price, cost and value), three principal valuation approaches (market, income and cost), four premises of value and six bases of value.

While IVS are usually applicable only where a valuation is stated as being based on IVS, in Australia the adoption of IVS by the API has the effect that IVS are binding on API members.

9.2.3.1 International valuation standards – structure

IVS comprises the following sections:

- Glossary – defining terms used in IVS which must then be read as having that meaning attributed by IVSC in the definition;
- IVS General Standards and
- IVS Asset Standards.

The IVS General Standards comprise:

- IVS100 *Valuation Framework* – addressing valuer principles (ethics, competency, etc.), valuation process quality control, compliance and effective date;
- IVS101 *Scope of Work* – addressing the terms of a letter of engagement for a valuation and a valuation review;
- IVS102 *Bases of Value* – addressing the six bases of value, entity-specific factors, synergies, assumptions, special assumptions, transaction costs and allocation of value;
- IVS103 *Valuation Approaches* – addressing the three principal valuation approaches being the market, income and cost approach;
- IVS104 *Data and Inputs* – addressing the selection and use of data as inputs in the valuation with the aim of maximising the use of relevant and observable data, with environmental, social and governance considerations also addressed;
- IVS105 *Valuation Models* – addressing the selection and use of valuation models in the valuation process, with an IVS compliant valuation requiring application of professional judgment (so excluding an automated valuation model) and
- IVS 106 *Documentation and Reporting* – addressing the documentation and reporting required that is sufficient to describe and provide transparency to the intended user on the valuation approach, valuation method(s), inputs, valuation models, professional judgment and resultant value such that the valuation is IVS compliant.

The IVS Asset Standards comprise eight standards, including the following of relevance to regulatory valuation:

- IVS300 *Plant, Equipment and Infrastructure* – specifying detailed requirements for the valuation of plant, equipment and infrastructure;
- IVS 400 *Real Property Interests* – specifying detailed requirements for the valuation of investment, owner-occupied and other property and
- IVS410 *Development Property* – specifying detailed requirements for the valuation of contemplated or uncompleted development projects.

For the purposes of regulatory valuation, being the preparation of valuations for inclusion in financial reporting, regard should be given to each of the above General Standards and Asset

Standards, with IVS101, IVS104 and IVS106 providing a sound basis upon which to undertake the valuation process.

9.2.3.2 Three key concepts

While the three key concepts of price, cost and value are defined in the Glossary (IVSC, 2024) as follows:

- Price – *the monetary or other consideration asked, offered or paid for an asset or to transfer a liability. Price and value may be different*;
- Cost – *the consideration or expenditure required to acquire or create an asset* and
- Value – *the valuer's quantitative conclusion on the results of a valuation process that is fully compliant with the requirements of IVS as of a valuation date*,

greater context was provided in an earlier edition of IVS (IVSC, 2013):

- *price* – being distinct from value which is a matter of opinion, price is a matter of fact representing the outcome of a contemplated or actual transaction:

  > Price is the amount asked, offered or paid for an asset.
  >
  > (IVSC, 2013)

  While price may reflect the characteristics of an individual vendor and an individual purchaser rather than a market as a whole, and so may not be generalisable across a market, multiple consistent observations of price in a series of transactions may be indicative of a market level;
- *cost* – like price, cost is conceptually distinguishable from value in IVSs, being also a matter of fact rather than a matter of opinion. Further, like price, cost is observable but that may be the extent of the useful information contributed with potentially limited insight provided into value:

  > Cost is the amount required to acquire or create an asset. When that asset has been acquired or created, its cost is a fact. Price is related to cost because the price paid for an asset becomes its cost to the buyer.
  >
  > (IVSC, 2013)

  Cost may reflect the characteristics of a particular party which may result in a different acquisition cost or cost to create than would arise from typical market participants and so may not be generalisable across a market; and
- *value* – within IVSs, value is either a judgment of the most probable price in exchange or of the economic benefits of ownership which broadly accord with the concepts of market value and investment value, respectively (IVSC, 2013).

For the purposes of regulatory valuation, being the preparation of valuations for inclusion in financial reporting, the key concept of greatest relevance is *value*.

9.2.3.3 Three principal valuation approaches

IVS 103 (IVSC, 2024) specifies three principal valuation approaches, being the cost approach, the income approach and the market approach:

- *cost approach* – grounded in economic theory, being based on the economic principle of 'substitution':

  > The cost approach provides an indication of value using the economic principle that a buyer will pay no more for an asset than the cost to obtain an asset of equal utility, whether by purchase or by construction, unless undue time, inconvenience, risk or other factors are involved.
  >
  > (IVSC, 2024)

  Usually applied to the valuation of properties for which there is no evidence of transaction prices or identifiable actual or notional income streams (such as town halls, museums and so forth), the method is simple to understand in principle but challenging to apply in practice. IVSs identify three valuation methods within the *cost approach*, being the replacement cost method, the reproduction cost method and the summation method;
- *income approach* – grounded in economic theory, being based on the economic principle of 'anticipation of benefits':

  > The income approach provides an indication of value by converting projected future cash flow to a single current value.
  >
  > (IVSC, 2024)

  The *income approach* conceptualises value as a single expression of future cashflows, simplistically masking a range of complex measurement issues. Within Australia, the commonly adopted valuation methods within the *income approach* are the capitalisation of income method and the discounted cash flow method; and
- *market approach* – grounded in economic theory, being based on the economic principle of 'price equilibrium':

  > The market approach provides an indication of value by comparing the asset and/or liability with identical or comparable (that is similar) asset and/or liability for which price information is available.
  >
  > (IVSC, 2024)

  The *market approach* is, therefore, dependent upon there being sales evidence of comparable assets, with the absence of such sales evidence rendering the approach inapplicable.

For the purposes of regulatory valuation, being the preparation of valuations for inclusion in financial reporting, one of these three approaches may be applicable in the determination of value.

9.2.3.4 Four premises of value

IVSs are based on four premises of value which are specified in IVS102 (IVSC, 2024):

- the premise of highest and best use, which is that use, from a participant (not entity) perspective, which would produce the highest value for the asset but which must be physically possible, financially feasible and legally allowable (consistent with IFRS 13);
- the premise of current use/existing use, which is the current use/existing use of the asset, liability or group of assets and/or liabilities and which may differ from highest and best use;

- the premise of orderly liquidation, which describes the value of a group of assets that could be realised in a liquidation sale, given a reasonable period of time to find a purchaser (or purchasers), with the seller being compelled to sell on an as-is, where-is basis; and
- the premise of forced sale, which describes circumstances where a seller is under compulsion to sell and that, as a consequence, a proper marketing period is not possible and buyers may not be able to undertake due diligence (IVSC, 2024).

For the purposes of regulatory valuation, being the preparation of valuations for inclusion in financial reporting, current use/existing use and highest and best use are each important premises that should be clearly ascertained and stated as they may have a significant impact on the assessment of value.

9.2.3.5 Six bases of value

IVS102 (IVSC, 2024) specifies six bases of value, being market value, market rent, equitable value, investment value/worth, synergistic value and liquidation value. For the purposes of regulatory valuation, being the preparation of valuations for inclusion in financial reporting, market value and market rent are of greatest relevance, while an understanding of investment value/worth is important such that it is not inadvertently applied:

- *market value* – being defined in IVS102 as:

  > Market value is the estimated amount for which an asset and/or liability should exchange on the valuation date between a willing buyer and a willing seller in an arm's length transaction, after proper marketing and where the parties had each acted knowledgably, prudently and without compulsion.
  >
  > (IVSC, 2024)

  Readers will note the similarity, but not the identicality, of the IVSC definition of market value to the definition of market value in *Spencer v Commonwealth* (1907) 5 CLR 418. However, for the purpose of regulatory valuation, it is the similarity of market value to the IFRS definition of fair value (as considered in Section 9.4) that is of greater relevance; and
- *market rent* – being defined as:

  > Market rent is the estimated amount for which an interest in real property should be leased on the valuation date between a willing lessor and a willing lessee on appropriate lease terms in an arm's length transaction, after proper marketing and where the parties had each acted knowledgably, prudently and without compulsion.
  >
  > (IVSC, 2024)

  For the purposes of regulatory valuation, *market rent* may be relevant for the valuation of owner-occupied property if the income approach is adopted and a notional rent is imputed for the property.

To avoid inadvertent application, *investment value* (often referred to in Commonwealth countries as worth) is defined in IVS 102 (IVSC, 2024) as:

> Investment value is the value of an asset to a particular owner or prospective owner for individual investment or operational objectives.
>
> (IVSC, 2024)

Care should be taken not to determine *investment value* for the purposes of regulatory valuation, as this is an assessment of value to a particular party, whereas *market value* and *fair value* are premised on value to hypothetical parties.

In summary, the above provides a heavily simplified guide to IVS for the purposes of regulatory valuation, being the preparation of valuations for inclusion in financial reports. While the provisions of IFRS and accounting standards take precedence, readers are encouraged to review the IVS in their entirety before undertaking a regulatory valuation.

9.3 Valuation professional body standards and guidance

While the provisions of IFRS, accounting standards and IVS take precedence, from a valuation viewpoint, for the purposes of regulatory valuation in Australia, publications by RICS and API are of greatest relevance.

9.3.1 RICS Red Book

The principal valuation standards publication by RICS is *RICS Valuation – Global Standards* (RICS, 2024), commonly referred to as the Red Book. First published in 1976, the Red Book is fully harmonised with IVS, and this book is based on the Red Book edition effective 31 January 2025, which comprises the following sections:

- Introduction – including the purpose, coverage and structure of the Red Book;
- Glossary – defining terms used in the Red Book, which must then be read as having that meaning attributed by RICS in the definition;
- Professional Standards – being mandatory (unless otherwise stated) for all RICS members providing written valuations:
  - PS1 Compliance with standards where a written valuation is provided;
  - PS2 Ethics, competency, objectivity and disclosures;
- Valuation technical and performance standards – containing specific, mandatory (unless otherwise stated) requirements and related implementation guidance directed to the provision of a valuation that is IVS compliant:
  - VPS1 Terms of engagement (scope of work);
  - VPS2 Bases of value, assumptions and special assumptions;
  - VPS3 Valuation approaches and methods;
  - VPS4 Inspections, investigations and records;
  - VPS5 Valuation models;
  - VPS6 Valuation reports and
- Valuation practice guidance applications – being advisory and not mandatory and providing further implementation guidance, including in the following specific areas of relevance to regulatory valuation:
  - VPGA1 Valuation for financial reporting;
  - VPGA4 Valuation of trade-related properties;
  - VPGA5 Valuation of plant and equipment (including infrastructure) and
  - VPGA8 Valuation of real property interests.

For the purposes of regulatory valuation, being the preparation of valuations for inclusion in financial reporting, readers are encouraged to review IFRS, accounting standards and the Red Book in their entirety before undertaking a regulatory valuation.

9.3.2 API guidance papers and protocols

API does not publish valuation standards but adopts IVS in full. Since 1 July 2021, API publishes Guidance Papers and Protocols, with Guidance Papers not being mandatory, serving as a guide and measure of acceptable professional practice and conduct by an API member (www.api.org.au).

While not mandatory, a departure from or non-compliance with a Guidance Paper must be stated in an API member's report or advice, including:

- the reason for the departure or non-compliance and
- the impact that the departure or non-compliance may have on the content of the report,

with members advised to seek legal and/or other advice before departing from practice recommended in a Guidance Paper.

Thirty-seven Guidance Papers are listed on the API website as of 21 March 2025, including the following of specific relevance to regulatory valuation:

- ANZVGP 102 Market value of property, plant and equipment in a business;
- ANZVGP 111 Valuation procedures – real property and
- AVGP 302 Valuations of real property, plant and equipment for use in Australian financial reports.

For the purposes of regulatory valuation, being the preparation of valuations for inclusion in financial reporting, readers are encouraged to review API Guidance Papers and Protocols in their entirety before undertaking a regulatory valuation.

9.4 Fair value

As noted in Section 9.2.1, the value of property for financial reporting is required to be stated at fair value, with IFRS 13 *Fair Value Measurement* (IFRS, 2025) providing a framework for the measurement of fair value and the requirements for disclosure in financial statements, based on the fundamental premise that fair value is a market-based measurement, not an entity-specific measurement.

IFRS 13 and AASB13 define fair value as:

> the price that would be received to sell an asset or paid to transfer a liability in an orderly transaction between market participants at the measurement date (para 9, IFRS 13)
>
> (IFRS, 2025)

Stewart (2022) notes that AASB13 describes what should be taken into account in the assessment of fair value, including:

- characteristics – like condition and location of the asset and restrictions, if any, on the sale or use of the asset;
- assumptions – those that market participants would use when pricing the asset or liability, assuming that market participants act in their economic best interest;
- purchase or sale price – the exit price under current market conditions (ie sale price), regardless of whether that price is directly observable or estimated using another valuation technique;
- market for the asset – principal (or most advantageous) market;
- use of the asset – highest and best use;

- techniques – valuation techniques to be used that are appropriate in the circumstances with three widely used valuation techniques to measure fair value being market, cost or income approaches; and
- inputs – shall maximise the use of relevant observable inputs and minimise the use of unobservable inputs.

It should be noted that, for the purposes of regulatory valuation, the definition of market value by the IVSC and by the Courts in *Spencer v Commonwealth* (1907) 5 CLR 418 is not as relevant as the definition of fair value in IFRS 13 and AASB13 which are paramount. However, the IVSC considers the definition of market value in IVS to be generally consistent with the definition of fair value in IFRS:

> The definition of fair value in IFRS is different from the above. The IVSB considers that the definitions of fair value in IFRS are generally consistent with market value. The definition and application of fair value under IFRS are discussed in IVS 300 Valuations for Financial Reporting.
>
> (IVSC, 2013, para 39, page 21)

> The commentary in IFRS 13 and, in particular, the reference to market participants, an orderly transaction, the transaction taking place in the principal or most advantageous market and to the highest and best use of the asset, make it clear that fair value under IFRSs is generally consistent with the concept of market value as defined and discussed in the IVS Framework. For most practical purposes, therefore, market value under IVS will meet the fair value measurement requirement under IFRS 13 subject to some specific assumptions required by the accounting standard such as stipulations as to the unit of account or ignoring restrictions on sale.
>
> (IVSC, 2013, IVS 300, para G2, pages 94–95)

Effectively, therefore, issues considered by a valuer in the determination of market value under IVS may be similarly considered by a valuer in the determination of fair value under IFRS for the purposes of regulatory valuation.

9.5 Valuation for financial reporting purposes

Valuation for financial reporting purposes will be considered through the following lenses:

9.5.1 international valuation standards, professional body standards and guidance;
9.5.2 owner-occupied property;
9.5.3 investment property;
9.5.4 development property and
9.5.5 value in use.

9.5.1 International valuation standards, professional body standards and guidance

International valuation standards, professional body standards and guidance will be considered through the following lenses:

9.5.1.1 valuations for financial reporting;
9.5.1.2 real property interests and
9.5.1.3 plant and equipment.

9.5.1.1 Valuations for financial reporting

While IVSs do not specifically address financial statements, VPGA 1 *Valuation for financial reporting* in the RICS Red Book (RICS, 2024) and the API's AVGP 302 *Valuations of real property, plant and equipment for use in Australian financial reports* (API, 2021a) are of direct relevance.

While VPGA 1 *Valuation for inclusion in financial statements* in the RICS Red Book is advisory and not mandatory, it alerts RICS members to relevant mandatory material elsewhere in the Red Book. VPGA 1 notes that the relevant basis of value will be fair value, reiterating the application of mandatory general requirements regarding ethical standards, member qualification and professional competence, including demonstrable freedom from bias.

VPGA 1 stresses the importance of clarity in the terms of engagement for valuations for financial reporting, which will include the accounting standards and basis of value to be adopted and the responsibilities of the client, such as for provision of information. Where information is supplied by the client, objective consideration and the application of professional scepticism should be adopted when forming a view on its credibility.

For reporting, VPGA 1 stresses the importance of clear and accurate conclusions of the valuation that are neither ambiguous nor misleading, encouraging fuller disclosure such as, for example, description of valuation approaches and methods used and valuation reasoning to assist understanding by the client and other intended users. Further, the report should include full details of the accounting standards and definition of the basis of value adopted.

The API's AVGP 302 *Valuations of real property, plant and equipment for use in Australian financial reports* is a guidance paper designed to serve as a guide and measure of acceptable professional practice and conduct for an API member. AVGP 302 provides context concerning accounting standards and IVSs, noting that market value as defined by the IVSC will generally be consistent with fair value as defined in accounting standards (para 3.4).

Concerning assets held for sale, AVGP 302 notes that these would normally be valued in a manner consistent with usual market valuation principles applied to surplus assets (para 4.4). Further, for owner-occupied assets, AVGP 302 notes that where there is continuation of owner occupation and the income approach is adopted, it is common practice to adopt notional market lease terms and conditions, including a market rental, with no allowance for vacancy or letting up, which should be detailed in the valuation report (para 4.5).

AVGP 302 notes that the same framework applies to valuations for financial reporting by not-for-profit entities, but that AASB 1049 *Whole of Government and General Government Sector Financial Reporting* applies for public sector financial statements, though this also requires the adoption of fair value.

In the event that a valuer is instructed to prepare a valuation for financial reporting purposes and subsequently to prepare a valuation for another purpose, AVGP 302 notes that care should be taken to avoid any conflict of interest between the reporting entity and the instructing party for the latter valuation (para 5.2).

Finally, AVGP 302 addresses liaison with auditors, being an idiosyncratic aspect of valuations for financial reporting purposes, including the requirement for assurance that the valuer's work constitutes appropriate audit evidence (such as sources of data, assumptions, methods and consistency with prior periods).

9.5.1.2 *Real property interests*

In addition to those professional body standards and guidance considered in section 9.5.1.1, the following also address real property interests, being of relevance for valuation for financial reporting purposes:

- IVS 400 *Real Property Interests* (IVSC, 2024);
- VPGA 8 *Valuation of real property interests* (RICS, 2024) and
- ANZVGP 111 *Valuation procedures* (API, 2021b) and ANZVGP 102 *Market value of property, plant and equipment in a business* (API, 2021c).

IVS 400 *Real Property Interests* supplements the General Standards and specifies additional requirements for the valuation of real property interests. The nature of the property interest to be valued should be clearly stated (superior interest (freehold), subordinate interest (leasehold) or right of use without exclusive possession) together with the extent of investigations undertaken and the nature and sources of information relied upon and any special assumptions agreed with the instructing party.

Selection of an appropriate basis of value, valuation approach and data inputs are addressed in IVS 400 together with the hierarchy of interests and rent. The hierarchy of interests concerns the ability for each interest in real property to have a different value – in an Australian context, this may include the freehold interest, head-leasehold interest, a range of sub-leasehold interests and possibly some licences or rights of use interests.

The valuer should identify the nature of the rights accruing to the holder of the interest being valued and reflect any constraints or encumbrances imposed by the existence of the other interests. Each interest will have a separate value and the sum of the separate values will frequently differ from the value of the freehold (being the unencumbered superior interest).

Concerning rent, when valuing the superior interest or the leasehold interest, the valuer must consider the contract rent payable under the lease as well as the market rent, which may be different.

VPGA 8 *Valuation of real property interests* in the RICS Red Book (RICS, 2024) is advisory and not mandatory, focusing on inspections and investigations with reference to sustainability and ESG. Issues that may only be apparent during an inspection, but that will impact on the market's perception of the value of a real property interest, are considered at length, such as the characteristics of the surrounding area and the accessibility of the property.

Investigations address the importance of information on title, building condition and services, planning and potential or actual constraints on the enjoyment of the property caused by sustainability and ESG factors and their impact on value. VPGA 8 notes that sustainability encompasses a wide range of physical, social, environmental and economic factors, including transition and stranding risk, that can affect value and of which valuers should be aware.

The API's ANZVGP 111 *Valuation procedures – real property* addresses instructions, report content, inspections and market evidence/analysis at length as well as valuation bases and considerations for owner-occupied property, being that the valuer should consider the purpose of the valuation (such as financial reporting) to determine the correct valuation basis and report qualifications.

Further, the API's ANZVGP 102 *Market value of property, plant and equipment in a business* provides guidance on valuation approaches where the valuation of property, plant and equipment forming part of a business is required.

9.5.1.3 Plant and equipment

In addition to those professional body standards and guidance considered in section 9.5.1.1, the valuation of plant and equipment is also addressed in:

- IVS300 *Plant and Equipment* (IVSC, 2024);
- VPGA 5 *Valuation of plant and equipment* in the RICS Red Book (RICS, 2024) and
- ANZVGP 102 *Market value of property, plant and equipment in a business* (API, 2021c).

IVS 300 *Plant and Equipment* notes that the valuation of plant and equipment requires consideration of the asset itself, its environment and its physical, functional and economic potential, such that an inspection should normally be undertaken. Clear assumptions are required concerning whether the plant and equipment will be valued as a whole, in place and as part of an operating business or on a different basis.

Further, the importance of selecting the basis of valuation is emphasised, as the valuation of plant and equipment is particularly sensitive to valuation in use, for orderly liquidation or for forced liquidation, with each of the valuation approaches considered in the context of plant and equipment. The possible impact of financing on value should be considered, as an asset subject to finance cannot be sold without the lender having been paid the balance outstanding under the financing arrangement. Items of plant and equipment that are subject to operating leases are the property of third parties and are therefore not included in a valuation of the assets of the lessee.

VPGA 5 *Valuation of plant and equipment (including infrastructure)* in the RICS Red Book (RICS, 2024) provides additional commentary on the practical application of IVS 300. VPGA 5 considers what comprises plant, machinery and equipment for the purposes of valuations for financial reporting, with care required to avoid omission or double counting with real property. Items usually included in the valuation of the property interest comprise items associated with the provision of services (such as gas, water, fire protection, security and so forth), structures (such as chimneys and railway tracks) and land improvements or site infrastructure (such as roadways and car parking). When valuing plant and equipment on the basis of market value (and, hence, fair value for financial reporting purposes), consideration should be given to an assumption that the assets will remain in place or are valued for removal.

The API's ANZVGP 102 *Market value of property, plant and equipment in a business* provides guidance on valuation approaches where the valuation of property, plant and equipment forming part of a business is required, with, in the context of plant and equipment, a focus on physical, functional and economic obsolescence.

9.5.2 Owner-occupied property

The principal accounting standard for consideration when valuing owner-occupied property is AASB 116 *Property, plant and equipment* (AASB, 2015). Owner-occupied property is defined in IAS 140 *Investment property* (AASB, 2019a) as:

> Owner-occupied property is property held (by the owner of by the lessee as a right of use asset) for use in the production or supply of goods or services or for administrative purposes.
>
> (IAS 40, para 5, page 1578)

Property, plant and equipment are defined in AASB 116 as:

> Property, plant and equipment are tangible items that:
>
> (a) are held for use in the production or supply of goods or services, for rental to others, or for administrative purposes; and
> (b) are expected to be used during more than one period.
>
> (AASB 116, para 6, page 6)

AASB 116 applies to all property, plant and equipment except:

- property, plant and equipment classified as held for sale in accordance with AASB 5 *Non-current assets held for sale and discontinued operations*; and
- biological assets related to agricultural activity, exploration and evaluation assets, mineral rights, mineral reserves and similar non-regenerative assets (AASB 116, para 3, page 5).

Under AASB 116, an item of property, plant and equipment will only be recognised if:

- it is probable that future economic benefits associated with the item will flow to the entity; and
- the cost of the item can be measured reliably (AASB 116, para 7, page 6).

AASB 116 requires general recognition of property, plant and equipment at cost at the time it is incurred, including acquisition or construction costs and subsequent costs of adding to, replacing part of or servicing an item of property, plant and equipment (AASB 116, para 10, page 6).

After initial recognition, AASB 116 requires an entity to choose between the cost model and the revaluation model and then apply that chosen model to an entire class of property, plant and equipment (AASB 116, para 29, page 9). This is the point at which an entity may choose to appoint a valuer to value, for example, a manufacturing property for financial reporting purposes but, having done so, will then have to value all other manufacturing properties, being then precluded from accounting for any manufacturing properties using the cost model.

The revaluation model under AASB 116 states that:

> After recognition as an asset, an item of property, plant and equipment whose fair value can be measured reliably shall be carried at a revalued amount, being its fair value at the date of revaluation less any subsequent accumulated deprecation and subsequent accumulated impairment losses.
>
> (AASB 116, para 31, page 10)

and further that:

> Revaluations shall be made with sufficient regularity to ensure that the carrying amount does not differ materially from that which would be determined using fair value at the end of the reporting period.
>
> (AASB 116, para 31, page 10)

It should be noted that AASB 116 requires the revaluation of an entire class of property, plant and equipment and not the selective revaluation of individual properties or items of plant and equipment (AASB 116, para 36 and 37, page 10).

Pragmatically, to avoid carrying assets at amounts that differ materially from fair value or selective valuation, entities commonly undertake a rolling programme (e.g. annual, 3-yearly or 5-yearly revaluations) of a class of property, plant and equipment.

When considering AASB 116, valuers should have regard to IVS 400 *Real Property Interests* and IVS 300 *Plant, equipment and infrastructure* and the API's AVGP 302 *Valuations of real property, plant and equipment for use in Australian financial reports*.

9.5.3 Investment property

Having regard to the massive investment property portfolios held by sovereign wealth funds (such as the Future Fund), REITs (such as Goodman and Dexus) and superannuation funds (such as UniSuper and Aware), the valuation of investment property is a huge source of business for the valuation profession, with the principal accounting standard for consideration when valuing investment property being AASB 140 *Investment property* (AASB, 2019a).

Investment property is defined in AASB 140 as:

> Investment property is property (land or a building – or part of a building – or both) held (by the owner or by the lessee as a right-of-use asset) to earn rentals or for capital appreciation, or both, rather than for:
>
> (a) use in the production or supply of goods or services or for administrative purposes, or
> (b) sale in the ordinary course of business.
>
> (AASB 140, para 5, page 5)

Significantly, investment property is not:

- property held for sale in the ordinary course of business or property under construction or development for such sale (AASB 140, para 9(a), page 6); nor
- owner-occupied property (AASB 140, para 9(c), page 6); nor
- property leased to another entity under a finance lease (AASB 140, para 9(e), page 6); nor
- biological assets related to an agricultural activity (AASB 140, para 4(a), page 5); nor
- mineral rights nor mineral reserves such as oil, natural gas or similar non-regenerative resources (AASB 140, para 4(b), page 5).

Further, for the purposes of IFRS, changes in the capital value of investment property are conceptually viewed as either income (capital value increase) or expense (capital value decrease). This has resulted in major investment property portfolio owners, such as REITs, reporting two profit figures – one being a compliance profit reflecting value changes and one excluding value changes and showing the underlying operational profit from the property portfolio.

AASB 140 requires that investment property be initially measured at cost in financial statements, including related transaction costs provided they are directly attributable expenditure (AASB 140, para 20, page 7). For sovereign wealth funds, REITs, superannuation funds and so forth, initial measurement at cost on acquisition may include the acquisition price, stamp duty, registration fees, legal fees, consultants' fees and so forth, which may be likely to result in a cumulative cost that exceeds the value of the property acquired at the time of acquisition.

Subsequent to initial measurement at cost, an entity is required to choose to adopt either the cost model or the revaluation model for all investment properties. However, AASB 140 requires all entities to measure the fair value of investment property for the purposes of either recognition in financial reporting (if the entity uses the revaluation model) or disclosure (if the entity uses the cost model). An entity is encouraged, but not required, to measure the fair value of investment property on the basis of a valuation by an independent valuer who holds a recognised and relevant professional qualification and who has recent experience in the location and category of the investment property being valued (AASB 140, para 32, pages 8–9).

When considering AASB 140, valuers should have regard to IVS 400 *Real Property Interests* and IVS 300 *Plant, Equipment and Infrastructure* and the API's AVGP 302 *Valuations of real property, plant and equipment for use in Australian financial reports.*

9.5.4 Development property

AASB 140 *Investment property* includes *property that is being constructed or developed for future use as an investment property* as an example of an investment property (AASB 140, para 8(e), page 6) (AASB, 2019a), further stating that:

> If an entity determines that the fair value of an investment property under construction is not reliably measurable but expects the fair value of the property to be reliably measurable when construction is complete, it shall measure that investment property under construction at cost until either its fair value becomes reliably measurable or construction is completed (whichever is earlier).
>
> (AASB 140, para 53, page 10)

IVS410 *Development Property* (IVSC, 2024) specifies detailed requirements for the valuation of contemplated or uncompleted development projects. Development properties are defined as interests where development is required to achieve the highest and best use or where improvements are either being contemplated or are in progress at the valuation date, including the construction of buildings, provision of infrastructure on undeveloped land, redevelopment of land or the improvement/alteration of existing buildings or structures (20.01).

The three main valuation approaches, market, income and cost, may be applicable for the valuation of development property, with the main valuation method, the residual method, being *a hybrid of the market approach, the income approach and the cost approach* (60.01).

IVS 410 notes that the residual method *indicates the residual amount after deducting all known or anticipated costs required to complete the development from the anticipated value of the project when completed after consideration of the risks associated with completion of the project. This is known as residual value* (100.03).

The residual value is highly sensitive to changes in the constituent variables, which IVS 410 addresses individually and at length, being:

- completed property value;
- construction costs;
- consultant's fees;
- statutory fees;
- marketing costs;
- timetable;
- finance costs;
- development profit (on both land and building);
- contingency and
- discount rate (100.08).

IVS410 notes that, for an existing asset, it is necessary to establish the suitability of the subject property for the proposed development, which may require information or reports from other specialists, with IVS410 identifying 16 issues that may require specific investigation (110.01).

9.5.5 Value in use

Stewart (2022) notes that the adoption of value in use, for the purposes of assessing impairments, is a common exception to the use of fair value as the basis of value, with value in use being a specialised interpretation of a type of income approach to valuation akin to the investment basis of value in IVS.

Impairments are addressed in AASB 136 *Impairment of assets* (AASB, 2019b), with the objective being to prescribe procedures that an entity applies in order to ensure that its assets are carried at no more than their recoverable amount. AASB 136 defines recoverable amount as the higher of an asset's fair value less costs of disposal and its value in use (AASB 138, para 6, page 7).

AASB 136 defines value in use as:

> Value in use is the present value of the future cash flows expected to be derived from an asset or cash-generating unit.
>
> (AASB 136, para 6, page 8)

The following elements are to be reflected in the calculation of an asset's value in use under AASB 136:

- an estimate of the future cash flows the entity expects to derive from the asset;
- expectations about possible variations in the amount or timing of those future cash flows;
- the time value of money, represented by the current market risk-free rate of interest;
- the price for bearing the uncertainty inherent in the asset; and
- other factors, such as illiquidity, that market participants would reflect in pricing the future cash flows the entity expects to derive from the asset.

(AASB 138, para 30, page 11)

through the following steps:

- estimating the future cash inflows and outflows to be derived from continuing use of the asset and from its ultimate disposal; and
- applying the appropriate discount rate to those future cash flows.

(AASB 136, para 31, page 11)

with extensive guidance provided on the cash flows (paras 33–54) and on the discount rate (paras 55–57).

If, and only if, the recoverable amount of an asset is less than its carrying amount, the carrying amount of the asset shall be reduced to its recoverable amount. That reduction is an impairment loss. (AASB136, para 59, page 14)

While a valuer may be instructed to perform an analysis of value-in-use under AASB 136, it is, essentially, a management assessment of worth to the entity within the provisions of AASB

136. Therefore, as a management assessment, Stewart (2022) cautions that valuers should consider whether:

- they are in a position to prepare a valuation report described as an estimate of value-in-use, when they are not management; and
- if they do prepare a report that estimates value-in-use, it might be better described as a limited scope valuation or a valuation calculation.

Consistent with Stewart's caution, AVGP 302 further cautions:

> Recoverable Amount may also be determined as based on an asset's Value In Use. This is a specially defined basis of value relevant only to impairment testing. This basis of value effectively represents the specific value of the asset as used by the current owner under certain limited conditions. These limiting conditions include the exclusion of cash flows arising from capacity expansions or other value improvements that have yet to be implemented by the owner. If requested to perform such a valuation, the Valuer should seriously consider whether they are able to assess the value that the current owner can derive, and if the Valuer concludes that they are able to assess this value, that they are familiar with the conditions for the calculation of this basis of value under AASB 136.
>
> (AVGP 302, para 4.3)

9.6 Summary and conclusions

Focusing on the key issues from a valuer's perspective, Chapters 1 and 2 introduced the concepts and principles of statutory valuation, with statutory valuation for acquisition considered in Chapters 3, 4, 5 and 6. Statutory valuation for rating and taxing was considered in Chapters 7 and 8, with regulatory valuation then considered in Chapters 9 and 10.

Chapter 1 introduced the principles of value in statutory valuation, including the concepts of market value, liberal/conservative estimates of value, highest and best use, potentiality and confirmation of a foresight. This provided the essential foundation principles of statutory valuation that are then applied in the following chapters on statutory valuation.

Chapter 2 considered the practice of statutory valuation including valuation methods with an emphasis on the comparative method of valuation, together with statutory valuation conventions such as the before and after method and the piecemeal method. This provided an understanding of the valuation methods and conventions that may be preferred for application in the following chapters on statutory valuation.

Statutory valuation for acquisition was considered in Chapters 3, 4, 5 and 6. Chapter 3 introduced the principles of statutory valuation for acquisition, including the statutory framework and process, with a focus on the concept of disregard of the scheme in compensation assessment and the distinction between concepts of value and worth. This provided an insight into concepts of fairness and equity in statutory valuation in the context of the disregard of compulsion.

Chapter 4 explored the market value bases of compensation in acquisition for which statutory valuation may be required, including market value of the interest and the related market value compensation concepts of severance and injurious affection. This provided a broad understanding of the heads of compensation for acquisition on a value basis that may arise in statutory valuation.

Chapter 5 considered non-market value forms of compensation, including special value, which is noted to rarely arise, payment for solace arising from loss of home which is heavily dependent on the exact wording of the facilitating statute and the concept of disturbance for costs and losses suffered as a result of the acquisition. This provided a broad understanding of the heads of compensation for acquisition that are not necessarily on a value basis but that may arise in statutory valuation.

Chapter 6 concluded the chapters considering statutory valuation for acquisition by examining compensation for other interests such as easements and mortgages, mineral rights and profits à prendre, strata title and substrata, businesses and the emerging statutory valuation areas of native title and of indigenous cultural loss. This provided an appreciation of the diversity of interests that may be subject to acquisition and so may require valuation for the assessment of compensation.

Statutory valuation for rating and taxing was considered in Chapters 7 and 8. Chapter 7 introduced the principles of land as the basis for taxation through land tax, stamp duty and rates, considering the general principles of the valuation of land for taxation using the New South Wales statutory framework as an example. This provided a general understanding of the often arcane and artificial basis of statutory valuation upon which government collects very significant amounts of taxation revenue.

Chapter 8 explored the practice of statutory valuation for rating and taxing through exemptions, concessions and allowances that have evolved over the last century, together with contemporary issues such as mass appraisal, automated valuation models and artificial intelligence, the impact of major disasters on valuation for rating and taxing and the litigation process, including the valuer as an expert witness. This provided an insight into the continued evolution of statutory valuation to address issues as they arise over time, highlighting the importance of a principles-based approach to statutory valuation.

Regulatory valuation is considered in chapters 9 and 10 in the context of the preparation of valuations for inclusion in financial reporting. This chapter distinguished regulatory valuation from statutory valuation through:

- an emphasis on accounting standards and valuation standards rather than statute and case law;
- an emphasis on fair value as defined in international financial reporting standards rather than value as defined in *Spencer v Commonwealth* (1907) 5 CLR 418 and
- an emphasis on valuation for financial reporting purposes rather than for compulsory acquisition or taxation purposes,

with the distinction of adopting a wide range of valuation approaches rather than an emphasis on the comparative method of valuation being considered in the next chapter.

This chapter explored the principles of regulatory valuation with regard to the interrelated roles of International Accounting Standards, International Financial Reporting Standards and International Valuation Standards, focusing on the primacy of fair value in international financial reporting standards relevant to the valuation of property, plant and equipment. This provided an understanding of the fundamental conceptual differences and bases between statutory valuation and regulatory valuation.

Harmonised with International Financial Reporting Standards and international and Australian accounting standards, International Valuation Standards are introduced and their structure explored with a focus on the role of IVS300 *Plant, Equipment and Infrastructure*, IVS 400 *Real Property Interests* and IVS410 *Development Property*. Also fully harmonised, the professional body standards and guidance provided in the RICS Red Book and API guidance papers and protocols for valuations for financial reporting purposes were identified and considered.

Given the centrality of fair value to regulatory valuation, the application of fair value for financial reporting purposes was considered and its alignment with the IVSC definition of market value noted. Further, valuation for financial reporting purposes generally and in the context of owner-occupied property, investment property and development property in the various standards was explored together with the management assessment of worth to the entity that comprises value-in-use.

Finally, chapter 10 concludes regulatory valuation by exploring the practice of valuation approaches and methods relevant for regulatory valuation such as the income approach and the cost approach and the various valuation methods there within. This provides an appreciation of the wider range of valuation approaches and methods available to the valuer for regulatory valuation, reflecting that, unlike statutory valuation, there is no requirement for defence in Court (negligence claims excepted).

References

AASB (2015) *AASB 116 property, plant and equipment*, AASB, Melbourne

AASB (2019a) *AASB 140 investment property*, AASB, Melbourne

AASB (2019b) *AASB 136 impairment of assets*, AASB, Melbourne

API (2021a) *AVGP 302 valuations of real property, plant and equipment for use in Australian financial reports*, API, Deakin

API (2021b) *ANZVGP 111 valuation procedures – real property*, API, Deakin

API (2021c) *ANZVGP 102 market value of property, plant and equipment in a business*, API, Deakin

IFRS (2025) *IFRS accounting standards 2025*, IFRS Foundation, London

IVSC (2013) *International valuation standards*, IVSC, London

IVSC (2024) *International valuation standards*, IVSC, London

Parker, D (2016) *International valuation standards – a guide to the valuation of real property assets*, Routledge, Abingdon

Parker, D (2022) Valuation principles, in D Parker (Ed) *Principles and practice of property valuation in Australia*, Routledge, Abingdon

Parker, D (2024) *Introduction to property valuation in Australia*, Routledge, Abingdon

RICS (2024) *RICS valuation – global standards (red book)*, RICS, London

Stewart, R (2022) Valuation for financial reporting purposes, in D Parker (Ed) *Principles and practice of property valuation in Australia*, Routledge, Abingdon

10 Regulatory valuation

Practice

This chapter draws heavily on the following, which are duly acknowledged here rather than individually referenced through the chapter and to which readers are referred for a deeper consideration:

Parker, D (2016) *International valuation standards – a guide to the valuation of real property assets*, Routledge, Abingdon
Parker, D (2022) Valuation principles, in D Parker (Ed) *Principles and practice of property valuation in Australia*, Routledge, Abingdon
Parker, D (2024) *Introduction to property valuation in Australia*, Routledge, Abingdon
Stewart, R (2022) Valuation for financial reporting purposes in D Parker (Ed) *Principles and practice of property valuation in Australia*, Routledge, Abingdon

10.1 Introduction

This book is an introduction to two very different valuation worlds in Australia – the complex and often arcane national realm of statutory valuation and the dynamic global empire of regulatory valuation. Statutory valuation is different from regulatory valuation, with the former being based on relevant applicable statutes with valuations being capable of challenge on appeal in Court while the latter is based on relevant applicable regulations with valuations incapable of challenge in Court (negligence claims excepted).

As an introductory textbook on statutory and regulatory valuation, this book is a companion to the Australian valuation textbooks, *Introduction to Property Valuation in Australia* (Parker, 2024) and *Principles and Practice of Property Valuation in Australia* (Parker, 2022), which provide a wider analysis of key principles underlying property valuation and current techniques and issues in the practice of property valuation for major sectors of the Australian property market.

Statutory valuation in Australia is most commonly encountered in valuations for rating, taxing and acquisition but may arise wherever a statute prescribes a basis of or approach to valuation. Regulatory valuation is most commonly encountered in valuations for financial statements and taxation purposes but may arise wherever a regulation prescribes a basis of or approach to valuation.

Rating and taxing of land raise significant amounts of revenue for State Governments and Councils in Australia, based on statutory valuation. For example, the NSW 2025–26 Budget Statement estimates land tax revenue for 2024–2025 to be $8.2 billion with the City of Brisbane Annual Plan and Budget 2024–2025 estimating proposed rates and utility charges to be $1.51 billion.

Acquisition of land, property and businesses commonly arises for major infrastructure projects and may be by agreement or by compulsion, based on statutory valuation. For example,

DOI: 10.1201/9781032616957-10

such major infrastructure projects in Australia currently include the multi-billion-dollar Sydney Metro, Melbourne Suburban Rail Loop and Melbourne to Brisbane Inland Rail projects.

Valuation of land and property holdings based on regulatory valuation arises for all reporting entities in Australia that publish financial statements upon which it is reasonable to expect that users will rely for decision making. For example, at June 2025 there are over 2,000 reporting entities listed on the Australian Stock Exchange which each publish financial statements, with the financial statements for Goodman Group, as at 30 June 2024, showing the fair value of total property assets to be $20.2 billion, based on regulatory valuation.

Focusing on the key issues from a valuer's perspective, the concepts and principles of statutory valuation are introduced in Chapters 1 and 2, with statutory valuation for acquisition considered in Chapters 3, 4, 5 and 6 and statutory valuation for rating and taxing considered in Chapters 7 and 8. Regulatory valuation is then considered in Chapters 9 and 10.

Chapter 1 introduced the principles of value in statutory valuation, including the concepts of market value, liberal/conservative estimates of value, highest and best use, potentiality and confirmation of a foresight. This provided the essential foundation principles of statutory valuation that are then applied in the following chapters on statutory valuation.

Chapter 2 considered the practice of statutory valuation including valuation methods for statutory valuation with an emphasis on the comparative method of valuation, together with statutory valuation conventions such as the before and after method and the piecemeal method. This provided an understanding of the valuation methods and conventions that may be preferred in the following chapters on statutory valuation.

Statutory valuation for land acquisition is considered in Chapters 3, 4, 5 and 6, supported by an evolving example in Appendix 1 of each chapter. Chapter 3 introduced the principles of statutory valuation for land acquisition, including the statutory framework and process, with a focus on the concept of disregard of the scheme in compensation assessment and the distinction between concepts of value and worth. This provided an insight into concepts of fairness and equity in statutory valuation in the context of the requirement to disregard the compulsory nature of the acquisition.

Chapter 4 explored the market value bases of compensation in acquisition for which statutory valuation may be required, including market value of the interest and the related market value compensation concepts of severance and injurious affection. This provided a broad understanding of the heads of compensation for acquisition on a value basis that may arise in statutory valuation.

Chapter 5 considered non-market value forms of compensation including special value, which was noted to rarely arise, payment for solace arising from loss of the primary place of residence and the concept of disturbance for costs and losses suffered as a result of the acquisition. This provided a broad understanding of the heads of compensation for acquisition that are not necessarily on a value basis but which may arise in statutory valuation.

Chapter 6 concluded the chapters considering statutory valuation for acquisition by examining compensation for other interests such as easements and mortgages, mineral rights and profits à prendre, strata title and substrata, businesses and the emerging statutory valuation areas of native title and indigenous cultural loss. This provided an appreciation of the diversity of interests that may be subject to acquisition and so may require valuation for the assessment of compensation.

Statutory valuation for rating and taxing was considered in Chapters 7 and 8. Chapter 7 introduced the principles of land as the basis for taxation through land tax, stamp duty and rates, considering the general principles of the valuation of land for taxation using the New South

Wales statutory framework as an example. This provided a general understanding of the often arcane and artificial basis of statutory valuation upon which government collects very significant amounts of taxation revenue.

Chapter 8 explored the practice of statutory valuation for rating and taxing through exemptions, concessions and allowances that have evolved over the last century, together with contemporary issues such as mass appraisal, automated valuation models and artificial intelligence, the impact of major disasters on valuation for rating and taxing and the litigation process including the valuer as an expert witness. This provided an insight into the continued evolution of statutory valuation to address issues as they arise over time, highlighting the importance of a principles-based approach to statutory valuation.

Regulatory valuation, considered in the context of the preparation of valuations for inclusion in financial reporting, is considered in Chapters 9 and 10. Regulatory valuation differs from statutory valuation in various ways including:

- an emphasis on accounting standards and valuation standards rather than statute and case law;
- an emphasis on fair value as defined in international financial reporting standards rather than value as defined in *Spencer v Commonwealth* (1907) 5 CLR 418;
- an emphasis on valuation for financial reporting purposes rather than for compulsory acquisition or land taxation purposes (considered in Chapter 9) and
- having a wide range of valuation approaches available for adoption (considered in this chapter) rather than an emphasis on the market approach and the comparative method of valuation.

Chapter 9 explored the principles of regulatory valuation with regard to the inter-related roles of international accounting standards, international financial reporting standards and international valuation standards, focusing on the primacy of fair value as defined in international financial reporting standards. This provided an understanding of the fundamental conceptual differences and bases between statutory valuation and regulatory valuation.

This chapter concludes regulatory valuation by exploring the practice of valuation approaches and methods relevant for regulatory valuation such as the income approach and the cost approach and the various valuation methods there within. This provides an appreciation of the wider range of valuation approaches and methods available to the valuer for regulatory valuation reflecting that, unlike statutory valuation, there is no requirement for defence of the valuation in Court (negligence claims excepted).

This book is based on statute, case law and regulations as at March 2025 and those standards and guidance documents published in IVSC (2024) and RICS (2024) and on the API website (accessed January to March 2025). Given their nature, statutes, case law, regulations, standards and guidance documents are dynamic and change continuously. Accordingly, readers should not rely on this book as a current statement of a statute, case law position, regulation, standard or guidance document and should undertake their own enquiries to find the most recent version.

While the market approach and the comparative method of valuation may be applied in regulatory valuation, these have been considered in Chapter 2. Accordingly, this chapter will focus on the income and cost approaches, considering:

- investment, owner-occupied and development property;
- income approach to valuation – capitalisation of income;
- income approach to valuation – discounted cash flow and
- cost approach to valuation.

10.2 Investment, owner-occupied and development property

Entities preparing financial statements may hold a wide range of property such as REITs owning office buildings and shopping centres which are leased to tenants, manufacturers owning production plants where they make goods for sale and property developers owning blocks of land for the purposes of current or future development.

Each such entity is required to reflect its ownership of such property assets in its financial statements and may elect to do so through periodic valuation. While there are various classifications for different property assets, for the purposes of regulatory valuation it is proposed to consider such assets in three groups, being investment property, owner-occupied property and development property.

10.2.1 Investment property

As noted in Section 9.5.3 of Chapter 9, the valuation of investment property is a huge source of business for the valuation profession with each of the shopping centres held by Scentre, each of the office properties held by Dexus and each of the logistics and distribution complexes held by Goodman, as well as each of the properties held by major superannuation funds such as ART, Aware and UniSuper, requiring valuation on a periodic basis for inclusion in financial statements.

The definition and accounting treatment of investment property is addressed by AASB 140 (Section 9.5.3, Chapter 9) which requires all entities to measure the fair value of investment property, with entities encouraged, but not required, to measure the fair value of investment property on the basis of a valuation by an independent valuer who holds a recognised and relevant professional qualification and who has recent experience in the location and category of the investment property being valued (AASB 140, para 32, pages 8–9) (AASB, 2019).

While IVSs do not specifically address financial reporting, VPGA 1 *Valuation for financial reporting* in the RICS Red Book (RICS, 2024) and the API's AVGP 302 *Valuations of real property, plant and equipment for use in Australian financial reports* (API, 2021a) are of direct relevance.

Further, the following international valuation standards, professional body standards and guidance address real property interests, being of relevance to valuers for valuation for financial reporting purposes:

- IVS 400 *Real Property Interests* (IVSC, 2024);
- IVS300 *Plant and Equipment* (IVSC, 2024);
- VPGA 5 *Valuation of plant and equipment* (RICS, 2024)
- VPGA 8 *Valuation of real property interests* (RICS, 2024) and
- ANZVGP 111 *Valuation procedures* (API, 2021b).

Following the definition of investment property in AASB 140 as:

> Investment property is property (land or a building – or part of a building – or both) held (by the owner or by the lessee as a right-of-use asset) to earn rentals or for capital appreciation, or both, rather than for:
>
> (a) use in the production or supply of goods or services or for administrative purposes, or
> (b) sale in the ordinary course of business.
>
> (AASB 140, para 5, page 5) (AASB, 2019)

the central feature is the existence of *rentals* or an income stream. Accordingly, investment property will generally be valued using an income approach to valuation through the capitalisation of income method, the profits method or the discounted cash flow method, as considered in Sections 10.3 and 10.4.

10.2.2 Owner-occupied property

While the valuation of owner-occupied property is a smaller source of business for the valuation profession than the valuation of investment property, each entity holding owner-occupied property may require valuations of such assets on a periodic basis for inclusion in financial statements.

As noted in Section 9.5.2 in Chapter 9, the principal accounting standard for consideration when valuing owner-occupied property is AASB 116 *Property, plant and equipment* (AASB, 2015) with owner-occupied property defined in AASB 140 *Investment property* (AASB, 2019) as:

> Owner-occupied property is property held (by the owner or by the lessee as a right of use asset) for use in the production or supply of goods or services or for administrative purposes.
>
> (AASB 140, para 5, page 1578)

While AASB 116 requires general recognition of property, plant and equipment at cost at the time it is incurred, after initial recognition an entity is required to choose between the cost model and the revaluation model and then apply that chosen model to an entire class of property, plant and equipment (AASB 116, para 29, page 9) (AASB, 2015). This is the point at which an entity may choose to appoint a valuer to value, for example, an owner-occupied warehouse property for financial reporting purposes but, having done so, will then have to value all other warehouse properties, being then precluded from accounting for any using the cost model.

While IVSs do not specifically address financial reporting, VPGA 1 *Valuation for financial reporting* in the RICS Red Book (RICS, 2024) and the API's AVGP 302 *Valuations of real property, plant and equipment for use in Australian financial reports* (API, 2021a) are of direct relevance.

Further, the following international valuation standards, professional body standards and guidance address real property interests, being of relevance for valuation for financial reporting purposes:

- IVS 400 *Real Property Interests* (IVSC, 2024);
- IVS300 *Plant and Equipment* (IVSC, 2024);
- VPGA 5 *Valuation of plant and equipment* (RICS, 2024)
- VPGA 8 *Valuation of real property interests* (RICS, 2024) and
- ANZVGP 111 *Valuation procedures* (API, 2021b).

Following the definition of owner-occupied property in AASB 140, there may be no rental or income stream passing with which to value the property. In the absence of an income stream, IVS 400 *Real Property Interests* (IVSC, 2024) states:

> in an owner-occupied building, it could be an assumed rent (or rent saved) based on what it would cost the owner to lease equivalent space.
>
> (80.01)

The valuer should, therefore, clearly state in the valuation report each of the assumptions made for such an *assumed rent*, including relevant assumed lease conditions such as lease length, rent review frequency and basis, liability for outgoings and so forth.

The valuer may choose to adopt *market rent*, as defined in IVS 102 *Bases of value* (IVSC, 2024):

> Market rent is the estimated amount for which an interest in real property should be leased on the valuation date between a willing lessor and a willing lessee on appropriate lease terms in an arm's-length transaction, after proper marketing and where the parties had each acted knowledgeably, prudently and without compulsion.
>
> (A20.01)

and clearly state the assumed *appropriate lease terms* in the valuation report. It should be noted that such *appropriate lease terms* are plural (so there are more than one) and may vary by sector (such as between a department store and a medical centre), by local market conventions and by local legislation (such as the impact of retail shop leases legislation in various States).

Further, concerning incentives, VPS 2 *Bases of value, assumptions and special assumptions* (RICS, 2024) states:

> Valuers must therefore set out clearly the principal lease terms that are assumed when providing an opinion of market rent. If it is the market norm for lettings to include a payment or concession by one party to the other as an incentive to enter into a lease, and this is reflected in the general level of rents agreed, the market rent should also be expressed on this basis. The nature of the incentive assumed must be stated by the valuer, along with the assumed lease terms.
>
> (5.3)

Therefore, the valuation of owner-occupied property usually requires the assessment of an income stream and valuation using an income approach to valuation through the capitalisation of income method or the discounted cash flow method, as considered in Sections 10.3 and 10.4

However, if the property is of a specialist nature where the physical structure is inextricably linked to the business that operates therein (such as a pub or cinema), then use of the profits method of valuation may be more appropriate.

Further, if the property is of a nature for which there is no active trading market or no identifiable income stream (such as a museum or church), then use of a cost approach may be more appropriate, such as replacement cost or reproduction cost, as considered in Section 10.5.

Valuers should also be cognisant of the role of value in use in the valuation of owner-occupied property for inclusion in financial reporting (Section 9.5.5, Chapter 9). Stewart (2022) notes that the adoption of value in use, for the purposes of assessing impairments, is a common exception to the use of fair value as the basis of value, with value in use being a specialised interpretation of a type of income approach to valuation akin to the investment basis of value in IVS.

10.2.3 Development property

AASB 140 *Investment property* (AASB, 2019) includes *property that is being constructed or developed for future use as an investment property* as an example of an investment property with IVS410 *Development Property* (IVSC, 2024) specifying detailed requirements for the valuation of contemplated or uncompleted development projects.

Development properties are defined, in IVS 410, as interests where development is required to achieve the highest and best use or where improvements are either being contemplated or are in progress at the valuation date, including the construction of buildings, provision of infrastructure on undeveloped land, redevelopment of land or the improvement/alteration of existing buildings or structures (20.01) (IVSC, 2024).

IVS 410 *Development Property* (IVSC, 2024) notes that, in the valuation of development property, regard should be given to the probability that any contracts in place (such as construction contracts or sale contracts) may become void or voidable in the event that one of the parties becomes insolvent and also to any contractual obligations that may have a material impact on market value.

Further, IVS 410 also notes that the valuation of development property often includes a significant number of assumptions and special assumptions regarding the condition or status of the project when complete. Key special assumptions may include the completion of construction and the leasing of the property, with variations in each having a potentially significant impact on the assessment of value.

IVS 410 further notes that, for an existing asset, it is necessary to establish the suitability of the subject property for the proposed development which may require information or reports from other specialists, with IVS 410 identifying 16 issues that may require specific investigation (110.01) (IVSC, 2024).

While some development property may be valued using the market approach or the income approach, it is common to use the static residual valuation method and dynamic hypothetical development method within the cost approach for the valuation of development property.

10.2.4 Summary – investment, owner-occupied and development property

Any of the market approach, income approach and cost approach may be applied in the valuation of different types of investment, owner-occupied and development property in the preparation of valuations for inclusion in financial statements.

While the market approach may be adopted for the valuation of investment, owner-occupied and development property, this approach has been addressed in Chapter 2. Table 10.1 sets out the other valuation approaches and valuation methods that may be used for the valuation of investment, owner-occupied and development property for inclusion in financial reporting which are considered further in Sections 10.3, 10.4 and 10.5.

Table 10.1 Property type, valuation approach and valuation method

Property type	*Valuation approach*	*Valuation method*
Investment property	Income approach	Capitalisation of income method Discounted cash flow method
Owner-occupied property	Income approach	Capitalisation of income method Profits method Discounted cash flow method
	Cost approach	Replacement cost method Reproduction cost method
Development property	Cost approach	Residual method Hypothetical development method

Source: Author

10.3 Income approach to valuation – capitalisation of income

The income approach to valuation is grounded in economic theory being based on the economic principle of anticipation of benefits. Within the income approach to valuation, the following methods may be adopted for the valuation of property for inclusion in financial statements:

Capitalisation of income method	Investment property and owner-occupied property
Profits method	Owner-occupied property

with each considered further below.

While the capitalisation of income method is considered below through the application to freehold interests in property, it may also be applied to long and short term leasehold interests which is considered in detail in Parker (2024).

10.3.1 Capitalisation of income – principles

IVS 105 *Valuation approaches* (IVSC, 2024) defines the income approach as follows:

> The income approach provides an indication of value by converting projected cash flows to a single current value. Under the income approach, the value of an asset is determined by reference to the value of income, cash flow or cost savings generated by the asset.
>
> (30.01)

VPS 3 *Valuation approaches and methods* (RICS, 2024) states:

> The income approach is based on capitalisation or conversion of present and predicted income (cash flows), which may take a number of different forms, to produce a single current capital value. Among the forms taken, capitalisation of a conventional market based income or discounting of a specific income projection can both be considered appropriate, depending on the type of asset and whether such an approach would be adopted by market participants. The income approach can also consider the value of an asset in terms of projected costs and cost savings, for example capital expenditure on retrofit.
>
> (2)

There are, therefore, essentially two valuation methods within the income approach, being conversion of future cash flow to a single value by either:

- the application of a multiplier to the cash flow, being the static income approach known as the capitalisation method, which is considered in this section; and
- discounting the future cash flow to a present value amount, being the dynamic income approach known as the discounted cash flow method, which is considered in section 10.4.

In its simplest form, the capitalisation of income method may be expressed as:

Representative single period income	$1,000,000
Capitalised in perpetuity at 5% (5/100)	20
Capital value	$20,000,000

10.3.1.1 Capitalisation of income method

As freehold property is capable of being held in perpetuity, the net income from the property is valued as being assumed to be receivable in perpetuity. Within the capitalisation of income method, there are two key elements for consideration, being the representative single period income and the capitalisation rate.

While deceptively simple, each variable masks a myriad of complex issues. The assessment of the representative single period income is heavily dependent on heuristics (being rules of thumb) and the determination of the capitalisation rate is heavily dependent on judgment, such that the reliability of the capitalisation of income method is inverse to the complexity of the investment property being valued. Further, the capitalisation of income method is most appropriate for adoption where there is a stable property in a stable property market with stable comparable property transactions and each having a relatively simple risk profile.

10.3.1.1.1 REPRESENTATIVE SINGLE PERIOD INCOME

In the absence of higher/lower rents under existing leases, the representative single period income is the market rent of the property as defined in IVS 102 *Bases of value* (IVSC, 2024) which is then capitalised in perpetuity. This requires the valuer to determine the market gross rent, deduct the assessed unrecoverable outgoings and so determine the net income for capitalisation with regard to transactions of comparable property in the relevant market.

To the determined net rent is applied a series of heuristics or rules of thumb including:

- assuming that the single period income will comprise the constant income;
- assuming that the single period income will comprise the income in perpetuity;
- assuming that the useful life of the property will be in perpetuity;
- making manually calculated income allowances for currently vacant space and/or space occupied at more/less that market rent;
- making percentage or similar adjustments for possible future vacancy and possible future incentives and
- linking future tenant retention and leasing strategies to an assessment of possible capital losses for deduction.

If the market rent is greater or lesser than the rental passing under the lease(s) for the property, the passing rental would be the representative single period income capitalised up to the date of the relevant rent review or lease expiry and the market rent would then be the representative single period income capitalised in perpetuity, deferred up to the date of the relevant rent review or lease expiry. It should be noted that the market rent is not escalated but capitalised at some point in the future based on current market levels. As such a property exhibits a reversion to market rent after a term of periods, this approach to the capitalisation of income method is conventionally referred to as a term and reversion valuation.

10.3.1.1.2 CAPITALISATION RATE

The capitalisation rate is the multiplier applied to the representative single period income to determine capital value. The capitalisation rate is determined based on comparable investment property sales and derived through adopting the accumulation, analysis, adjustment and application process considered in Section 2.2.1 of Chapter 2.

The adjustment step requires the use of significant professional judgment, as all of the differences between two comparable properties (location, building, tenant, etc.) are captured in one adjustment which may be very small – for example, 5.0% to 5.25% – but which can have significant implications for the assessment of capital value.

Where the passing rent is higher or lower than the market rent or there is a gap in the net income stream due to vacancy/releasing or an incentive, the valuation will need to reflect the term and reversion nature of the cash flow.

For the term, the calculation of the present value of a cash flow for a given number of periods at a given rate of interest may be calculated using the present value of $1pa formula which may be expressed as:

$$PV\$1pa = \frac{1 - \frac{1}{(1+i)^n}}{i}$$

with

PV$1pa	being the present value of the future cash flow
i	being the given rate of interest
n	being the given number of periods of the cash flow

Therefore, the multiplier for a cash flow for 5 years at 5% will be 4.3295.

For the reversion, the calculation of a right to receive a cash flow in perpetuity after a given number of periods at a given rate of interest is usually expressed as *PV $1 for n years at i%,* being calculated using the present value formula which may be expressed as:

$$PV\$1 = \frac{1}{(1+i)^n}$$

with

PV$	being the present value of the future cash flow
i	being the given rate of interest
n	being the given number of periods of the cash flow

Therefore, the multiplier for the present value of a cash flow in perpetuity which is available after 5 years at 5%, usually expressed as *PV 5 years at 5%,* will be 0.7835. This multiplier is then applied to the value of the cash flow in perpetuity to determine the present value of the reversion, which is then added to the value of the term to determine the value of the investment property.

While these formulae may be easily calculated using a calculator or Excel, *Parry's Valuation and Conversion Tables* (Davidson, 1978) have provided ready reckoner tables of such multipliers for over a century.

Conventionally, the assessment of capital value is only rounded off once at the end of the valuation calculations with no rounding undertaken at each of the stages in the valuation process (such as the calculation of the term and the reversion, above). The level of rounding adopted should reflect that adopted by market participants.

For further information on the principles of the capitalisation of income method, see Parker (2024).

10.3.1.2 Profits method

The profits method is generally adopted where the type of property is inextricably linked to the business that operates therein, such as a hotel, pub, bar, restaurant, nightclub, petrol filling station, casino, cinema, theatre and various other forms of leisure property. Such properties are often owned rather than leased so that evidence of leasing transactions may be limited and are generally sold fully fitted out as operational entities such that evidence of sales requires careful analysis. Accordingly, the profits method seeks to examine the economic fundamentals of the business as the source of value of the property by capitalising the estimated future trading potential rather than net income (Wyatt, 2023).

IVS400 *Real Property Interests* (IVSC, 2024) notes that, when adopting the profits method of valuation, the valuer should also comply with the requirements of IVS200 *Business and business interests* and IVS210 *Intangible assets* (80.3). IVS200 *Business and business interests* (IVSC, 2024) requires that the capitalisation or discount rate applied must be consistent with the type of income or cash flow used which may be on a pre-tax, post-tax or other basis (60.03). Further, in estimating the appropriate capitalisation rate or discount rate, factors such as the level of interest rates, rates of return expected by participants for similar investments and the risk inherent in the anticipated benefit stream are to be considered (60.06) (Parker, 2024).

VPGA4 *Valuation of individual trade related properties* in the RICS Red Book (RICS, 2024) provides advisory (not mandatory) guidance on the profits method of valuation when applied to the valuation of an individual property on the basis of trading potential (1.2). The profits method applies to property that is designed or adapted for a specific use and the resulting lack of flexibility usually means that the value of the property interest is intrinsically linked to the returns that the owner can generate from that use. The value therefore reflects the trading potential of the property which may be contrasted with generic property (such as office, retail or industrial) that can be occupied by a range of different business types (1.3). (Parker, 2024)

Wyatt (2023) helpfully provides further clarification of the profits method as follows:

Estimate	Sales (fair maintainable turnover – FMT) – note that actual trading accounts may not be relevant
Deduct	Cost of sales (purchases and adjustment for change in value of stock)
=	Gross profit
Deduct	Operating costs (energy costs, rates, insurance, maintenance, marketing, etc.) and wage costs
=	Net profit before remuneration to operator/proprietor
Deduct	Arms' length remuneration to operator/proprietor
Deduct	Interest on capital invested, stock and consumables
=	Fair maintainable operating profit (FMOP)
Then either	(1) capital valuation – capitalise FMOP at an appropriate freehold or leasehold capitalisation rate; or
	(2) rental valuation – apportion FMOP between rent and profits (the divisible balance)

The profits method is a relatively simple approach to a complex valuation with Wyatt (2023) noting the following simplifying premises:

- an assumption that the business will at all times be effectively and competently managed, operated and promoted and that it is properly staffed, stocked and capitalised;
- that the valuation includes land, buildings, trade fixtures, fittings, furnishings and equipment associated with the business and assumes they are working and owned outright;
- the inclusion of market perception of inherent trading potential including locational goodwill (attached to the property);
- an assumption that advance bookings and order books can be transferred;

- an assumption that existing licences, consents, permits, certificates and registrations can be obtained and renewed;
- an assumption (usually) that trade inventory is included;
- an assumption that personal goodwill is excluded; and
- an assumption that wet and dry consumable stock and any badged items and the value attached to brand name should be separately identified (page 197).

Baum et al (2018) note that the process of undertaking a profits method valuation, following the steps outlined earlier, involves a significant amount of detail and so may be very time consuming. While shorthand measures may be used as a cross check, such as rent/value per cinema seat, restaurant cover, barrel of beer sold, litre of petrol sold and so forth, the consistent analysis of comparable sales in imperfect submarkets is challenging.

It should be noted that valuers who prepare valuations of trade related property will usually specialise in a particular market, such as theatres and cinemas or hotels, pubs and bars, as knowledge of the operational aspects of the property and of the industry as a whole are fundamental to understanding the market transactions and the analysis required.

For further information on the principles of the profits method, see Parker (2024).

10.3.2 Capitalisation of income – practice

Within the income approach to the valuation of investment property and owner-occupied property, the capitalisation of income and profits methods of valuation may be adopted for valuations for inclusion in financial statements, with the following providing worked examples of each.

10.3.2.1 Capitalisation of income method

You are instructed to assess the market value of the following properties for inclusion in financial statements:

- Property A – an outer suburban office park of 15,000sqm NLA which is leased to Tenant X for a term of five years for $200psm pa, being the market rent, with no rent reviews and with three years of the lease term unexpired – being a rack-rented or fully rented investment; and
- Property B – an inner suburban office building of 1,000sqm NLA which is leased to Tenant Y for $75psm pa, being below market rent, for a term of five years with three years of the lease term unexpired, following which there may be an increase in rental which is known as a reversion – being a reversionary investment.

10.3.2.1.1 PROPERTY A – RACK-RENTED OR FULLY RENTED INVESTMENT

Your accumulation, analysis, adjustment and application of comparable market rentals in the outer suburban office park sub-market indicate a current market rental value of $200psm pa to be applicable to the subject property, being a rack-rented or fully rented investment.

Your analysis of the lease to Tenant X finds that the tenant pays all outgoings, so no deduction for unrecoverable outgoings is required.

Your accumulation, analysis, adjustment and application of comparable sales transactions and discussions with investors active in the outer suburban office park sub-market lead you to conclude that a capitalisation rate of 6% is applicable to the subject property.

In this example, the representative single period income is the passing rental, which is the same as the market rental, being $200 psm pa or $3,000,000 pa. As the tenant pays all outgoings, no deduction for unrecoverable outgoings is required.

Further, as the passing rental is the same as the market rental, there is no reversionary income.

The applicable capitalisation rate is 6% and there are no gaps in income or voids, so the capitalisation of income valuation would be undertaken as follows:

$$\text{Value} = \frac{\text{Market rent}}{\text{Capitalisation rate}} = \frac{\$3,000,000}{0.06} = \$50,000,000$$

Alternatively, the Years Purchase or YP may be applied being the reciprocal of the capitalisation rate and an expression of the number of years of net income required to recoup the present value or outlay:

Net income	$3,000,000
x YP in perpetuity at 6%	16.6667
Valuation	$50,000,100

As the valuation is not a round number, rounding off is required which, in this example, would be rounding to $50 million.

10.3.2.1.2 PROPERTY B – REVERSIONARY INVESTMENT

Your accumulation, analysis, adjustment and application of comparable market rentals in the inner suburban office property sub-market indicate a current market rental value of $100psm pa to be applicable to the subject property.

Your analysis of the lease to Tenant Y finds that the tenant pays all outgoings, so no deduction for unrecoverable outgoings is required.

Your accumulation, analysis, adjustment and application of comparable sales transactions and discussions with investors active in the inner suburban office property sub-market lead you to conclude that a capitalisation rate of 7.5% is applicable to rack-rented investments like the subject property.

In this example, the representative single period income is the market rental, being $100 psm pa or $100,000 pa. As the tenant pays all outgoings, no deduction for unrecoverable outgoings is required.

The reversionary income is the increase from lease rental to market rental ($75 psm pa to $100 psm pa or $75,000 pa to $100,000 pa) which is attainable at the end of the lease in three years' time.

The applicable capitalisation rate is 7.5% for a rack-rented investment and there are no gaps in income or voids. Conventionally, the term income is considered safer and the reversionary income is considered riskier (particularly if there is a long time period before the reversion is realised), reflected by a change in the capitalisation rate which is usually based on valuer judgment rather than market evidence.

The capitalisation of income valuation would be undertaken as follows:

Term		
Term rent passing	$75,000	
YP 3 years at 7.0%	2.6243	$196,822
Reversion:		
Reversion to market rent	$100,000	
YP perpetuity at 7.5%	13.3333	
Deferred 3 years (PV $1 for 3 years at 7.5%)	0.804960	$1,073,277
Valuation		$1,270,099

As the valuation is not a round number, rounding off is required which, in this example, would be rounding to $1.27 million.

For further information on the practice of the capitalisation of income method, see Parker (2024).

10.3.2.2 Profits method

You are instructed to assess the market value of the freehold interest in a country motel with 30 bedrooms, bar and restaurant for inclusion in financial statements. You have been provided with the last five years accounts for the motel which you consider to be a reasonable estimate of fair maintainable turnover (FMT) for the accommodation, bar and restaurant and allow you to form an estimate of costs of sale, operating costs and wage costs together with remuneration to the operator and interest on capital invested, stock and consumables.

Your accumulation, analysis, adjustment and application of comparable sales transactions and discussions with investors and operators active in the country motel market lead you to conclude that a capitalisation rate of 15% is applicable to the subject property.

Following Wyatt (2023), the profits method of valuation would be undertaken as follows:

Estimate	Sales (fair maintainable turnover – FMT)	Accommodation	$1,000,000
		Bar	$100,000
		Restaurant	$25,000
		Total	$1,125,000
Deduct	Cost of sales (purchases and adjustment for change in value of stock)	Purchases	($80,000)
		Adj for depreciation in value of stock	($45,000)
		Total	($125,000)
=	Gross profit		$1,000,000
Deduct	Operating costs (energy costs, rates, insurance, maintenance, marketing, etc.) and wage costs	Operating	($100,000)
		Wages	($150,000)
		Total	($250,000)
=	Net profit before remuneration to operator/proprietor		$750,000
Deduct	Arm's length remuneration to operator/proprietor		($100,000)
Deduct	Interest on capital invested, stock and consumables	Furniture, fixtures, fittings, equipment	($350,000)
		Stock (avg)	($150,000)
		Cash-1 month working capital	($50,000)
		Operator's capital	($550,000)
		Annualised at 10%	($55,000)
=	Fair maintainable operating profit (FMOP)		$595,000
Capital valuation	Capitalise FMOP at an appropriate capitalisation rate	YP in perpetuity at 15%	$3,966,666

As the valuation is not a round number, rounding off is required which, in this example, would be rounding to $4.00 million.

The above example is a very simplified version of a complex valuation task which requires the valuer to have a detailed knowledge of both the relevant industry being valued and the operation of businesses within that industry. Unless the valuer is experienced in the valuation of properties within the specific industry, such as hotels, pubs, bars, restaurants, nightclubs and various other forms of leisure property, then assistance should be sought from a valuer expert in that specific industry (with client consent in the scope of works and acknowledgment in the valuation report).

For further information on the practice of the profits method, see Parker (2024).

10.4 Income approach to valuation – discounted cash flow

The income approach to valuation is grounded in economic theory being based on the economic principle of anticipation of benefits. Within the income approach to valuation, the discounted cash flow method may be adopted for the valuation of investment property and owner-occupied property for inclusion in financial reporting, as considered further below.

10.4.1 Discounted cash flow – principles

As noted in Section 10.3.1, there are, essentially, two valuation methods within the income approach, being conversion of future cash flow to a single value by either:

- the application of a multiplier to the cash flow being the static income approach known as the capitalisation method, which was considered in Section 10.3; and
- discounting the future cash flow to a present value amount being the dynamic income approach known as the discounted cash flow method, which is considered in this section.

IVS 103 *Valuation approaches* (IVSC, 2024) states:

> Under the DCF method the forecasted cash flow is discounted back to the valuation date, resulting in a present value of the asset.
>
> (A20.02)
>
> In some circumstances for long-lived or indefinite-lived assets, DCF may include a terminal value which represents the value of the asset at the end of the explicit projection period. In other circumstances, the value of an asset may be calculated solely using a terminal value with no explicit projection period. This is sometimes referred to as an income capitalisation method.
>
> (A20.03)
>
> The key steps in the DCF method are:
>
> (a) choose the most appropriate type of cash flow for the nature of the subject asset and the valuation (i.e. pre-tax or post-tax, total cash flows or cash flows to equity, real or nominal, etc),
> (b) determine the most appropriate explicit period, if any, over which the cash flow will be forecast,
> (c) prepare cash flow forecasts for that period,
> (d) determine whether a terminal value is appropriate for the subject asset at the end of the explicit forecast period (if any) and then determine the appropriate terminal value for the nature of the asset,
> (e) determine the appropriate discount rate, and
> (f) apply the discount rate to the forecasted future cash flow, including the terminal value, if any.
>
> (A20.04)

The most appropriate type of cash flow is considered in IVS 103 *Valuation approaches* (IVSC, 2024) with alternatives including whole or partial interest cash flow, pre-tax or post-tax cash flow, nominal or real cash flow, different currencies and varying forms such as contractual cash flow, most likely cash flow, expected cash flow and so forth (A20.05). Importantly, the discount rate should be assessed on the same basis as the cash flow.

The most appropriate explicit period over which the cash flow will be forecast is considered in IVS 103 *Valuation approaches* (IVSC, 2024). In determining the period, regard should be given to the life of the asset, a reasonable period for which reliable data is available on which to base projections and a minimum explicit forecast period which should be sufficient for an asset to achieve a stabilised level of growth, after which a terminal value can be used (A20.09).

In Australia, the market value of investment property will usually be determined using the nominal whole cash flow for the asset in Australian dollars on a pre-tax basis for a ten-year period, disregarding financing, being contractual cash flows for known periods and forecasts for the balance.

While the income from a freehold property is assumed to be received in perpetuity, the value of that income stream beyond year 10 of a 10-year cash flow is conventionally represented by an assessment of the terminal value of the property (IVS 103, A20.25) (IVSC, 2024). The terminal value is usually assessed by assuming a sale of the property on the last day of year 10, with the buyer being willing to pay for future cashflows starting in year 11. Hence, the terminal value is usually assessed by capitalising the year 11 net income which is then included in year 10 of the cash flow and discounted back at the discount rate to contribute to the present value.

This necessitates the assessment of a terminal capitalisation rate which may be the current applicable capitalisation rate less an allowance for obsolescence, for example an increase in the capitalisation rate of 25–50 bps.

Concerning the discount rate, IVS 103 *Valuation approaches* (IVSC, 2024) states:

> The rate at which the forecast cash flow is discounted should reflect not only the time value of money, but also the risks associated with the type of cash flow and the future operations of the asset.
>
> (A20.29)

While the preferred basis for assessment of the discount rate is by reference to comparable sales, a build-up approach comprising a risk free rate (such as the Commonwealth Government 10 year bond rate) plus a risk premium may be adopted or the discount rate may be derived through a capital asset pricing model or the weighted average cost of capital adopted (IVS 103, A20.31). Importantly, the discount rate should be assessed on the same basis as the cash flow.

For further information on the principles of the discounted cash flow method, see Parker (2024).

10.4.2 Discounted cash flow – practice

Within the income approach to the valuation of investment property, the discounted cash flow method of valuation may be adopted for valuation for inclusion in financial statements. The worked example, below, following Parker (2024), applies the discounted cash flow method of valuation for the purposes of illustration only and the data therein should not be adopted unchanged by readers undertaking a discounted cash flow valuation of another property.

You are instructed to assess the market value of the freehold interest in a logistics park for inclusion in financial statements. The logistics park is 10 years old and comprises four buildings, each leased to a separate tenant:

Building	*NLA sqm*	*Passing rent*	*Rent review*	*Lease expiry*
A	20,000 sqm	$300 psm	5 yearly market rent review in year 4 and 9	3 years after end year 10
B	10,000 sqm	$400 psm	No rent review	End year 3
C	15,000 sqm	$350 psm	No rent review	End year 5
D	5,000 sqm	$450 psm	No rent review	End year 4

Your accumulation, analysis, adjustment and application of comparable market rentals in the logistics park sub-sector indicate a current open market rental value of:

Building A	$350 psm pa
Building B	$450 psm pa
Building C	$400 psm pa
Building D	$500 psm pa

to be applicable to the subject property with forecast rental growth of 3% pa.

Your analysis of outgoings costs in the subject property and recovery provisions under the leases indicate an annual unrecoverable outgoings cost of $50 psm pa which will grow by the annual rate of inflation which a consensus of economists has assessed at 2.5%pa.

Your analysis of the logistics park sub-sector outlook leads you to estimate that it will take 3 months to relet the space occupied by Tenant B and Tenant D for a term of 10 years each with a market rent review after 5 years, but 6 months to relet the space occupied by Tenant C for a term of 5 years, with reletting costs being approximately 10% of the first year's rent for each.

Given the imminent lease expiries following the end of the cash flow in 10 years' time, you consider an upgrade of the park roadways and parking areas, in an attempt to retain tenants, will be required at an estimated cost of $5,000,000.

Your accumulation, analysis, adjustment and application of comparable sales transactions and discussions with investors active in the logistics park sub-sector lead you to conclude that a discount rate of 8.5% is applicable to the subject property and that an appropriate capitalisation rate for the subject property would be in the order of 5.5%.

Table 10.2 comprises a DCF in a simple Excel spreadsheet format. Time is on the horizontal axis rows from Year 0 to Year 11 and the key variables are on the vertical axis columns comprising:

- gross cash flow, tenant by tenant, summed;
- deduction for unrecovered outgoings;
- deduction for reletting costs;
- deduction for capex;
- inclusion of terminal value and
- net cash flow.

The open market rental growth is calculated at the top of the Excel spreadsheet, being the current open market rental value for each building with forecast rental growth of 3% pa.

This is a highly simplified DCF Excel spreadsheet and readers may expect to include greater complexity in their valuation spreadsheets to reflect the complexity of the relevant property being valued and it's property market sub-sector.

Table 10.2 Discounted cash flow valuation method – example

1	A	B	C	D	E	F	G	H	I	J	K	L	M	N	O
2															
3				**Year 0**	**Year 1**	**Year 2**	**Year 3**	**Year 4**	**Year 5**	**Year 6**	**Year 7**	**Year 8**	**Year 9**	**Year 10**	**Year 11**
4	Market rental value psm:														
5	Building A	20,000 sqm			$350	$361	$371	$382	$394	$406	$418	$430	$443	$457	$470
6	Building B	10,000 sqm			$450	$464	$477	$492	$506	$522	$537	$553	$570	$587	$605
7	Building C	15,000 sqm			$400	$412	$424	$437	$450	$464	$478	$492	$507	$522	$538
8	Building D	5,000 sqm			$500	$515	$530	$546	$563	$580	$597	$615	$633	$652	$672
9															
10	Passing rental:														
11	Building A	20,000 sqm			$6,000,000	$6,000,000	$6,000,000	$7,649,089	$7,649,089	$7,649,089	$7,649,089	$7,649,089	$8,867,391	$8,867,391	$8,867,391
12	Building B	10,000 sqm			$4,000,000	$4,000,000	$4,000,000	$3,687,954	$4,917,272	$4,917,272	$4,917,272	$4,917,272	$5,700,465	$5,700,465	$5,700,465
13	Building C	15,000 sqm			$5,250,000	$5,250,000	$5,250,000	$5,250,000	$5,250,000	$3,477,822	$6,955,644	$6,955,644	$6,955,644	$6,955,644	$6,955,644
14	Building D	5,000 sqm			$2,250,000	$2,250,000	$2,250,000	$2,250,000	$2,110,329	$2,813,772	$2,813,772	$2,813,772	$2,813,772	$3,261,933	$3,261,933
15	**Gross Cash Flow**				**$17,500,000**	**$17,500,000**	**$17,500,000**	**$18,837,043**	**$19,926,690**	**$18,857,955**	**$22,335,777**	**$22,335,777**	**$24,337,272**	**$24,785,433**	**$24,785,433**
16															
17	Less unrecovered outgoings				-$2,500,000	-$2,562,500	-$2,626,563	-$2,692,227	-$2,759,532	-$2,828,521	-$2,899,234	-$2,971,714	-$3,046,007	-$3,122,157	-$3,200,211
18	Less reletting costs:														
19		Tenant B						-$491,725							
20		Tenant C								-$695,562					
21		Tenant D							-$281,375						
22	Less capex													-$5,000,000	
23	Terminal value													$388,980,759	
24															
25	**Net Cash Flow**			**$0**	**$15,000,000**	**$14,937,500**	**$14,873,438**	**$15,653,091**	**$16,885,782**	**$15,333,872**	**$19,436,543**	**$19,364,063**	**$21,291,265**	**$405,644,035**	**$21,585,222**
26															
27	**PV**														
28	PV at 8.5%	$280,771,836		$0	$15,000,000	$14,937,500	$14,873,438	$15,653,091	$16,885,782	$15,333,872	$19,436,543	$19,364,063	$21,291,265	$405,644,035	
29															
30	**NPV**														
31	Net Cash Flow			-$280,000,000	$15,000,000	$14,937,500	$14,873,438	$15,653,091	$16,885,782	$15,333,872	$19,436,543	$19,364,063	$21,291,265	$405,644,035	
32	NPV at 8.5%	$711,370													
33															
34	**IRR**														
35	Net Cash Flow			-$280,000,000	$15,000,000	$14,937,500	$14,873,438	$15,653,091	$16,885,782	$15,333,872	$19,436,543	$19,364,063	$21,291,265	$405,644,035	
36	IRR	8.54%													
37															
38															

Source: Author

10.4.2.1 Type of cash flow

Consistent with the usual approach adopted in Australia, the value of investment property using DCF will be determined using the nominal whole cash flow for the asset in Australian dollars on a pre-tax basis, being contractual cash flows under the leases for known periods and forecast cash flows for the balance.

10.4.2.2 Explicit forecast period

Consistent with the usual approach adopted in Australia, the value of investment property using DCF will be determined using a ten-year forecast period. Year 0 is included to allow for inclusion of the acquisition cost of the property in the net present value (NPV) and internal rate of return (IRR) calculations, with Year 11 included to facilitate the calculation of the terminal value.

10.4.2.3 Cash flow forecasts

In this simplified example, the cash flow forecasts have been limited to:

- gross cash flow, tenant by tenant, summed;
- deduction for unrecovered outgoings;
- deduction for reletting costs;
- deduction for capex;
- inclusion of terminal value and
- net cash flow.

10.4.2.3.1 GROSS CASH FLOW, TENANT BY TENANT, SUMMED

For each tenant, the contractual cash flow is included up to rent review or lease expiry. In the case of the rent review for Tenant A, the rent is increased to market level at rent review in Year 4 and Year 9. In the case of Tenants B, C and D, following lease expiry there is a loss of rent during the reletting period followed by an increase to market rent at the start of the new lease. For simplicity, incentives and outgoings lost due to vacancy have been disregarded.

10.4.2.3.2 DEDUCTION FOR UNRECOVERED OUTGOINGS

Tenants are assumed to contribute to the cost of some outgoings, leaving the owner with a shortfall for unrecovered outgoings. Your analysis of outgoings costs in the subject property and recovery provisions under the leases indicate an annual unrecoverable outgoings cost of $50 psm pa which will grow by the annual rate of inflation which a consensus of economists has assessed at 2.5% pa.

10.4.2.3.3 DEDUCTION FOR RELETTING COSTS

Your analysis of the logistics park sub-sector outlook leads you to estimate that reletting costs will be approximately 10% of the first year's rent for each tenant, which is included as a deduction in the Excel spreadsheet.

10.4.2.3.4 DEDUCTION FOR CAPEX

Allowances for capital expenditure may be for known items (such as air-conditioning replacement) or for anticipated items (such as refurbishment). As each of the leases in the subject property will expire soon after the end of Year 10, there is a significant potential future vacancy risk which may be reflected either in the capitalisation rate adopted in the terminal value and/or in an allowance for capital expenditure to upgrade the subject property in the hope of retaining tenants or finding new tenants more quickly should they vacate at lease expiry (with due caution to avoid double counting).

You have determined that an upgrade of the logistics park roadways and parking areas should be allowed in the cash flow for Year 10 and your discussions with architects, designers, engineers and builders suggest that an allowance of $5,000,000 would be appropriate. It should be noted that such an allowance should be borne in mind by the valuer when considering the capitalisation rate in the assessment of terminal value, with due caution to avoid double counting.

10.4.2.3.5 NET CASH FLOW

Net cash flow is net income from the property, being the sum of gross cash flow less deductions for unrecovered outgoings, reletting costs and capex. When terminal value is included, Years 1 to 10 in the net cash flow row are the amounts discounted to find the present value of the subject property.

10.4.2.4 Terminal value

The terminal value is calculated by capitalising the Year 11 net cash flow by an appropriate capitalisation rate, adjusting for any reversions or vacancies and including the terminal value in Year 10, effectively assuming a sale of the subject property on the last day of Year 10 based on the Year 11 cash flow.

Your analysis of recent sales transactions and discussions with investors active in the logistics park sub-sector lead you to conclude that an appropriate capitalisation rate for the subject property in the current market would be in the order of 5.5%. Conventionally, allowing for deprecation and obsolescence, a 0.5% adjustment may be made to the capitalisation rate adopted to determine terminal value for this type of property which would result in 6.0% for the subject property.

However, as a significant allowance of $5,000,000 for an upgrade of the logistics park roadways and parking areas has been allowed in the net cash flow, your experience suggests that depreciation and obsolescence and possible tenant risk may be reduced such that you retain a capitalisation rate of 5.5% for the determination of terminal value and include this in the Year 10 cash flow, thus also avoiding double counting.

10.4.2.5 Discount rate

Your analysis of recent sales transactions and discussions with investors active in the logistics park sub-sector lead you to conclude that an observed or inferred discount rate of 8.5% is applicable to the subject property.

Your analysis and discussions also lead you to conclude that an appropriate capitalisation rate for the subject property in the current market would be in the order of 5.5% and your analysis of market rentals in the logistics park sub-sector indicate forecast rental growth of 3%pa. Applying the build-up method of the capitalisation rate (5.5%) plus rental growth rate (3%pa) indicates a discount rate of 8.5%, which supports the observed or inferred discount rate.

With the prevailing 10 year Commonwealth Government bond rate being 4%, applying the build-up method of risk free rate plus risk premium suggests that if the discount rate is 8.5%, the risk premium is 4.5%. Your analysis of recent sales transactions and discussions with investors active in the logistics park sub-sector lead you to conclude that this is a reasonable risk premium for the subject property.

10.4.2.6 Present value

Applying the Excel function to derive present value using a discount rate of 8.5% to the net cash flow for Years 1 to 10, =NPV(0.085,E28:N28), provides an assessment of present value of $280,771,836 rounded to $280 million.

10.4.2.7 Net present value

The calculation of net present value inserts the rounded assessment of present value in Year 0 and then applies the Excel function to derive net present value to the net cash flow for Years 0 to 10, =NPV(0.085,D31:N31), providing an assessment of net present value of $711,370.

In the example, the net present value is marginally positive, reflecting that the present value was rounded down rather than up in the assessment of value, indicating that acquisition at the assessed value will provide a return marginally above 8.5%. If a different acquisition cost were to be adopted, a different net present value would arise which may be positive or negative.

10.4.2.8 Internal rate of return

The calculation of internal rate of return inserts the rounded assessment of present value in Year 0 and then applies the Excel function to derive internal rate of return to the net cash flow for Years 0 to 10, =IRR(D35:N35), providing an internal rate of return of 8.54%.

In the example, the internal rate of return is very slightly above the discount rate (target rate), reflecting that the present value was rounded down rather than up in the assessment of value. Acquisition at $280 million will allow the investor to exceed the discount rate (target rate). If a different acquisition cost were to be adopted, a different internal rate of return would arise which may be above or below the discount rate (target rate).

For further information on the practice of the discounted cash flow method, see Parker (2024).

10.5 Cost approach to valuation

The cost approach to valuation is grounded in economic theory, being based on the economic principle that a buyer will pay no more than the cost to obtain an asset of equal utility, whether by purchase or construction.

Within the cost approach, replacement cost and reproduction cost methods may be adopted for valuation for inclusion in financial reporting of assets for which there is no active trading market or no identifiable income stream, such as a museum or church. The residual method and hypothetical development method may be adopted for valuation for inclusion in financial reporting of development assets, such as vacant land held for development or existing property held for redevelopment.

It should be noted, however, that the cost approach to valuation is fundamentally flawed by its combination of notions of cost and notions of value. Wyatt (2023) validly cautions against the careless intermingling of concepts of cost and concepts of value, with cost being a production-related concept and value being an exchange or use-related concept.

Parker (2024) observes that mixing notions of value and notions of cost is akin to trying to mix oil and water. Notions of value and notions of cost do not sit well together and the result is a curious expression that is neither a pure expression of value nor a pure expression of cost.

10.5.1 Cost methods – principles

Within the cost approach to valuation, the following methods may be adopted for the valuation of property for inclusion in financial statements:

Replacement cost method	Owner-occupied property
Reproduction cost method	Owner-occupied property
Residual method	Development property
Hypothetical development method	Development property

with each considered further below.

10.5.1.1 Replacement cost method

Replacement cost is based on the cost of replacing the utility of the asset, not the exact physical properties of the asset (IVS103, A30.02), adjusted for physical deterioration and relevant forms of obsolescence to provide depreciated replacement cost (IVS103, A30.03) (IVSC, 2024).

Parker (2024) notes that, in Australia today, depreciation should reflect physical, functional and economic obsolescence with typical depreciation profiles including:

- straight line, being the same proportion of the original cost deducted for each period of the estimated life of the asset; or
- diminishing value, being the deduction of a constant percentage rate from the cost at the start of the previous period over the estimated life of the asset; or
- S-curve, being the deduction of different percentage rates for each period over the estimated life of the asset, such as higher rates in the initial period reducing over the estimated life of the asset.

Hence, the replacement cost method of valuation may also be referred to as the depreciated replacement cost method of valuation (DRC method) or, if the building design is optimised to replicate the function and utility of the asset, the optimised depreciated replacement cost method of valuation (ODRC method).

Significantly, IVS103 (IVSC, 2024) notes that the replacement cost is generally that of a modern equivalent asset, being one that provides similar function and equivalent utility to the asset being valued, but which is of a current design and constructed or made using cost-effective materials and techniques (A30.05).

By way of example, replacement cost for the Sydney Opera House may be based on the cost of replacing an opera theatre, drama theatre, playhouse, concert hall, restaurant and car park in a current design using cost-effective materials and construction techniques. Accordingly, the replacement cost for the Sydney Opera House would be unlikely to include the internationally recognised convex 'sails' structure or current configuration of facilities (Parker, 2024).

In its simplest form, the replacement cost method may be expressed as:

	Cost of land
plus	Cost of a modern equivalent building
=	Total cost of modern equivalent property
less	Depreciation allowance on cost of modern equivalent building
=	Value of existing property

For further information on the principles of the replacement cost method, see Parker (2024).

10.5.1.2 Reproduction cost method

Reproduction cost is based on the cost of recreating a replica of the subject asset, adjusted for depreciation based on physical, functional and economic obsolescence associated with the subject asset.

Reproduction rather than replacement may be appropriate in circumstances where:

- the cost of a modern equivalent asset is greater than the cost of recreating a replica of the subject asset or
- the utility offered by the subject asset could only be provided by a replica rather than a modern equivalent (IVS103, 30.06) (IVSC, 2024).

For example, if an asset's exact design and features were an integral part of the benefit that would accrue to an owner, the equivalent utility could only be provided by a reproduction of the subject asset. An example would be an iconic building where the design was of greater importance than the functionality of the accommodation within it, such as the Sydney Opera House, where reproduction cost may be based on the cost of recreating a replica of the existing building including the internationally recognised convex 'sails' structure, adjusted for deprecation (Parker, 2024).

In its simplest form, the reproduction cost method may be expressed as:

	Cost of land
plus	Cost of a replica of the building
=	Total cost of a replica of the property
less	Depreciation allowance on cost of building
=	Value of existing property

For further information on the principles of the reproduction cost method, see Parker (2024).

10.5.1.3 Residual method

The residual method of valuation is, effectively, a static form of the hypothetical development method of valuation, being a single period statement of the present value of each input. The dynamic form of the hypothetical development method of valuation, being a multi-period cash flow approach usually calculated in a spreadsheet or proprietorial software system, is considered in Section 10.5.1.4.

The critical difference between the static and dynamic forms concerns the cost of finance, with the static approach relying on heuristic assumptions or rules of thumb while the dynamic approach sets out the cash flow in periods over the development allowing a more accurate assessment of the cost of finance and resulting amount available for land acquisition.

IVS 410 *Development property* (IVSC, 2024) states:

> The residual method is so called because it indicates the residual amount after deducting all known or anticipated costs required to complete the development from the anticipated value of the project when completed after consideration of the risks associated with completion of the project. This is known as the residual value.
>
> (100.03)

In its simplest form, the residual method when applied to determine the value of development land may be expressed as:

	Value of the completed development project
less	Total cost of the development project including developer's profit
=	Value of development land in present condition (residual value) (after Millington, 2000)

In practice, the residual method of valuation usually includes a much wider range of input variables which Wyatt (2023) summarises as follows:

Value of the completed development project – development value:	
	Floor area sqm
x	Market rental \$psmpa
=	Estimated rental value
x	Net initial yield or capitalisation rate
=	Gross development value before sale costs
-	Sale costs
=	Net development value after sale costs
Total cost of the development project – development costs:	
	Site preparation
+	Building costs – \$psm × area
+	External works
+	Professional fees – % building costs and external works
+	Contingency allowance – % building costs
+	Letting agents fees – % market rental
+	Letting legal fees – % market rental
=	Development costs
+	Developer's profit on development costs – %
+	Finance on development costs for half building period (being a heuristic assumption or rule of thumb)
+	Finance on development costs and finance for void period
=	Total development costs
Net development value less total development costs	
Land costs (calculated in reverse order):	
-	Finance on land costs over total development period
-	Developer's profit on land costs – %
=	Residual land value before purchase costs
-	Purchase costs
=	Value of development land in present condition – residual land value

As will be apparent, there are two fundamental flaws in the residual method of valuation:

- it combines notions of value and notions of cost and
- it requires separate assessments of a wide range of interconnected input variables with a small change in one variable not only affecting the assessment of residual land value but also impacting other variables which may then also change and further affect the assessment of residual land value.

Indeed, Rost and Collins (1984) succinctly note:

> Because of the many estimates and assumptions which must be made, the use of this method can be perilous.
>
> (page 65)

Accordingly, IVS 410 *Development property* (IVSC, 2024) cautions:

> The residual value can be highly sensitive to relatively small changes in the forecast cash flows and the practitioner should provide separate sensitivity analyses for each significant factor.
>
> (100.04)
>
> Caution is required in the use of this method because of the sensitivity of the result to changes in many of the inputs, which may not be precisely known on the valuation date, and therefore have to be estimated with the use of assumptions.
>
> (100.05)

Given that the residual value is highly sensitive to changes in the constituent variables, IVS 410 *Development property* (IVSC, 2024) addresses each of the key variables individually and at length:

- completed property value;
- construction costs;
- consultant's fees;
- statutory fees;
- marketing costs;
- timetable;
- finance costs;
- development profit (on both land and building);
- contingency and
- discount rate (100.08)

to which readers are referred for a more detailed consideration.

For further information on the principles of the residual method, see Parker (2024).

10.5.1.4 Hypothetical development method

The residual method of valuation is, effectively, a static form of the hypothetical development method of valuation, being a single period statement of the present value of each input, considered in Section 10.5.1.3. The dynamic form of the hypothetical development method of valuation, being a multi-period cash flow approach usually calculated in a spreadsheet or proprietorial software system, is considered below.

The critical difference between the static and dynamic forms concerns the cost of finance, with the static approach relying on heuristic assumptions or rules of thumb while the dynamic approach sets out the cash flow in periods over the development allowing a more accurate assessment of the cost of finance and resulting amount available for land acquisition.

The key inputs for the hypothetical development method are the same as for the residual method and may be either:

- inputs measured at the time that the cash flow is constructed which are assumed to remain constant for the duration of the cash flow period – such as the value of the completed development based on the market rental value prevailing at the time that the cash flow is constructed or
- inputs that are escalated over the duration of the cash flow to reflect an estimate of their cost/value at that point in the future when they are incurred – such as the value of the completed development based on the final period market rental value which has been escalated at a rental growth rate over the duration of the cash flow.

While it may be tempting for the valuer to use escalators within the hypothetical development method cash flow in an attempt to more closely reflect future costs/values, this adds a further set of assumptions to an already highly subjective valuation process which may be challenging to support.

For further information on the principles of the hypothetical development method, see Parker (2024).

10.5.2 Cost methods – practice

Within the cost approach to the valuation of owner-occupied property and development property, the replacement cost method, reproduction cost method, residual method and hypothetical development method of valuation may be adopted for valuations for inclusion in financial reporting, with the following providing worked examples of each.

10.5.2.1 Replacement cost method

For the valuation of owner-occupied property, the replacement cost method may be appropriate for valuations for inclusion in financial statements where there is no active trading market or no identifiable income stream, with replacement cost being based on the depreciated cost of replacing the utility of the asset not the exact physical properties of the asset.

You are instructed to assess the depreciated replacement cost of the freehold interest in an 500sqm art gallery located within a former inner suburban bank, built in 1895 with an elaborate stone façade but not heritage listed, for inclusion in financial statements.

The building has 70 years of its estimated 200-year life remaining. Construction costs for a modern art gallery providing equivalent utility are $4,000 psm according to Rawlinsons (2024) and the construction period is estimated to be 1 year, with professional fees estimated to be 10% of building costs.

Your accumulation, analysis, adjustment and application of comparable land sales in the inner suburban sub-market indicate a land value of $10,000 psm to be applicable to the gallery site of 1,000 sqm. The entity which owns the gallery can borrow funds for construction and land acquisition at an interest rate of 7.5% pa.

Following Wyatt (2023), the depreciated replacement cost may be calculated as follows:

Building:		
	Building area sqm	500 sqm
x	Building cost $psm	$4,000 psm
=	Cost of modern building of equivalent utility	$2,000,000
+	Fees, at 10% of build cost	$200,000
+	Finance costs for build cost and fees over half build period at 7.5%	$82,500
=	Gross replacement cost of modern building of equivalent utility	$2,282,500
x	Deprecation allowance 65% (age/estimated economic life assuming straight line depreciation)	($1,483,625)
=	Net replacement cost of modern building of equivalent utility	$798,875
Land:		
	Size of site sqm/hectares	1,000 sqm
x	Land value psm/per hectare	$10,000 psm
=	Value of land	$10,000,000
+	Finance costs for land over build period at 7.5%	$750,000
=	Cost of land	$10,750,000
Replacement cost valuation:		
=	Net replacement cost of modern building of equivalent utility	$798,875
+	Cost of land	$10,750,000
=		$11,548,875

As the valuation is not a round number, rounding off is required which, in this example, would be rounding to $11.50 million as the value of the art gallery for inclusion in financial statements.

Readers should note that the relativity of net replacement cost to land value highlights the issues in the cost approach concerning mixing concepts of cost and value, with the example also illustrating the impact of the building cost rate and depreciation period chosen on the output of the cost method. The relativity of net replacement cost to land value also raises issues of highest and best use and development potential, which are conveniently disregarded in the example.

For further information on the practice of the replacement cost method, see Parker (2024).

10.5.2.2 Reproduction cost method

For the valuation of owner-occupied property, the reproduction cost method may be appropriate for valuations for inclusion in financial statements where there is no active trading market or no identifiable income stream with reproduction cost being based on the depreciated cost of recreating a replica of the subject asset rather than replacing the utility of the asset.

You are instructed to assess the reproduction cost of a 2,000 sqm destination motor museum in a purpose built iconic building, designed by a famous Australian architect and heritage listed, in a small country town within easy driving distance of a State capital city, for inclusion in financial reporting.

The motor museum has 100 years of its estimated 150-year life remaining. Construction costs for a reproduction motor museum are $8,000 psm according to Rawlinsons (2024) and the construction period is estimated to be 1 year, with professional fees estimated to be 10% of building costs.

Your accumulation, analysis, adjustment and application of comparable land sales in the country town indicate a land value of $600 psm to be applicable to the motor museum site of 5,000 sqm. The entity, which owns the motor museum, can borrow funds for construction and land acquisition at an interest rate of 7.5% pa

Following Wyatt (2023), the reproduction cost may be calculated as follows:

Building:		
	Building area sqm	2,000 sqm
x	Building cost $psm	$8,000 psm
=	Cost of reproduction building	$16,000,000
+	Fees, at 10% of build cost	$1,600,000
+	Finance costs for build cost and fees over half build period at 7.5%	$660,000
=	Gross reproduction cost of building	$18,260,000
x	Deprecation allowance 33% (age/estimated economic life using straight line depreciation)	($6,025,800)
=	Net reproduction cost of building	$12,234,200
Land:		
	Size of site sqm/hectares	5,000 sqm
x	Land value psm per hectare	$600 psm
=	Value of land	$3,000,000
+	Finance costs for land over build period at 7.5%	$225,000
=	Cost of land	$3,225,000
Reproduction cost valuation:		
=	Net reproduction cost of building	$12,234,200
+	Cost of land	$3,225,000
=		$15,459,200

As the valuation is not a round number, rounding off is required which, in this example, would be rounding to $15.50 million as the value of the motor museum for inclusion in financial statements.

Readers should note that the relativity of net replacement cost to land value highlights the issues in the cost approach concerning mixing concepts of cost and value, with the example also illustrating the impact of the building cost rate and depreciation period chosen on the output of the cost method.

Further, readers should consider the usefulness of a reproduction cost valuation of $15.50 million, for a motor museum in a small country town, to persons seeking to make decisions based on the financial statements.

For further information on the practice of the reproduction cost method, see Parker (2024).

10.5.2.3 Residual method

For the valuation of development property, the residual method may be appropriate for valuations for inclusion in financial statements. The residual method of valuation is, effectively, a static form of the hypothetical development method of valuation, being a single period statement of the present value of each input. The dynamic form of the hypothetical development method of valuation, being a multi-period cash flow approach usually calculated in a spreadsheet or proprietorial software system, is considered in Section 10.5.2.4.

You are instructed to assess the market value of an outer suburban industrial development site for inclusion in financial statements. The site is 5,000 sqm upon which may be constructed an industrial building of 4,000 sqm gross floor area.

Your accumulation, analysis, adjustment and application of comparable market rentals in the outer suburban industrial property sub-market indicate a current market rental value of $250 psm pa to be applicable to the subject property, with the tenant paying all outgoings.

Your accumulation, analysis, adjustment and application of comparable sales transactions and discussions with investors active in the outer suburban industrial sub-market lead you to conclude that a capitalisation rate of 6.5% is applicable to the subject property.

Construction costs for an outer suburban industrial building are $1,400 psm according to Rawlinsons (2024) and the construction period is estimated to be 2 years, with the following estimates of costs and fees:

- sale costs at $100,000;
- site preparation costs at $25,000;
- external works at $75,000;
- professional fees at 10% of building costs;
- contingency allowance at 10% of building costs;
- finance available at an interest rate of 7.5% pa;
- letting agent's fees at 10% of market rental;
- letting legal fees at 5% of market rental and
- land purchase costs at 5%

with a developer anticipated to seek a profit margin of 15% on development costs.

Following Wyatt (2023), the residual method may be applied as shown in Table 10.3.

As the valuation is not a round number, rounding off is required which, in this example, would be rounding to $4.85 million as the value of the development site for inclusion in financial statements.

For further information on the practice of the residual method, see Parker (2024).

10.5.2.4 Hypothetical development method

For the valuation of development property, the hypothetical development method may be appropriate for valuations for inclusion in financial statements. While the residual method of valuation is, effectively, a static form of the hypothetical development method of valuation, being a single period statement of the present value of each input, considered in Section 10.5.2.3, the dynamic form of the hypothetical development method of valuation is a multi-period cash flow approach usually calculated in a spreadsheet or proprietorial software system.

The critical difference between the static and dynamic forms concerns the cost of finance, with the static approach relying on heuristic assumptions while the dynamic approach sets out the cash flow in periods over the development allowing a more accurate assessment of the cost of finance and resulting amount available for land acquisition.

Table 10.3 Residual valuation method – example

	Value of the completed development project – development value:			
	Floor area sqm	4,000 sqm		
x	Market rental $psmpa	$250 psm pa		
=	Estimated rental value		$1,000,000	
x	Net initial yield or capitalisation rate	6.5%	YP 15.384615	
=	Gross development value before sale costs			$15,384,615
-	Sale costs		$100,000	
=	Net development value after sale costs			$15,284,615
	Total cost of the development project – development costs:			
	Site preparation		$25,000	
+	Building costs – $psm × area	4,000 sqm × $1,400 psm	$5,600,000	
+	External works		$75,000	
+	Professional fees – % building costs and external	10%	$560,000	
+	Contingency allowance – % building costs	10%	$560,000	
+	Letting agents fees – % market rental	10%	$100,000	
+	Letting legal fees – % market rental	5%	$50,000	
=	Development costs			$6,970,000
+	Developer's profit on development costs – %	15%	$1,045,500	
+	Finance on development costs for half building period of 2 years	7.5% pa	$522,750	
=	Total development costs			$8,538,250
	Net development value less total development costs:			$6,746,365
	Land costs:			
-	Finance on land costs over total development period	PV 7.5% over 2 years (0.8653)	($908,736)	$5,837,629
-	Developer's profit on land costs – %	15%	($761,429)	
=	Residual land value before purchase costs			$5,076,200
-	Purchase costs	5%	($241,723)	
=	**Value of development land in present condition – residual land value**			$4,834,477

Source: Author

The example set out in Section 10.5.2.3 for the static form (residual method) is used below for the dynamic cash flow form of the hypothetical development method which, following Wyatt (2023), may be modelled simplistically in Excel as follows:

- the development period of 2 years is modelled as 8 quarters;
- finance is available at an interest rate of 1.875% per quarter;

- the net development value and development costs are spread over the 8 quarters to reflect their likely timing of being incurred, summed to a net cash flow before finance per quarter and
- the rolling cost of finance is modelled separately, being cumulative over the 8 quarters, as shown in Table 10.4.

Readers will note that a positive balance of $107,144 may be found in cell M22, being the amount by which the net development value exceeds development costs plus cost of finance. While this positive balance is impacted by the use of an interest rate of 1.875% per quarter rather than 7.5%pa, the key influence is the impact of the timing of cash flows in the dynamic cash flow approach compared to the heuristic assumptions of the static approach concerning:

- finance on development costs for half building period; and
- finance on land costs over total development period.

Recasting the model to eliminate the positive balance of $107,144 by iteration results in a residual land value of $4,912,358 which may be rounded to $4.90 million as the value of the development site for inclusion in financial statements.

In addition to modelling in Excel, there are a range of software providers currently operating in Australia, including Forbury, ARGUS, CoStar and Yardi, who provide sophisticated software packages for hypothetical development method valuation which are widely used and allow for very prescriptive modelling of the timing of cash flows and the finance costs arising.

For further information on the practice of the hypothetical development method, see Parker (2024).

10.6 Summary and conclusions

Focusing on the key issues from a valuer's perspective, Chapters 1 and 2 introduced the concepts and principles of statutory valuation, with statutory valuation for acquisition considered in chapters 3, 4, 5 and 6. Statutory valuation for rating and taxing was considered in chapters 7 and 8, with regulatory valuation then considered in chapters 9 and 10.

Chapter 1 introduced the principles of value in statutory valuation, including the concepts of market value, liberal/conservative estimates of value, highest and best use, potentiality and confirmation of a foresight. This provided the essential foundation principles of statutory valuation that are then applied in the following chapters on statutory valuation.

Chapter 2 considered the practice of statutory valuation including valuation methods with an emphasis on the comparative method of valuation, together with statutory valuation conventions such as the before and after method and the piecemeal method. This provided an understanding of the valuation methods and conventions that may be preferred for application in the following chapters on statutory valuation.

Statutory valuation for acquisition was considered in Chapters 3, 4, 5 and 6. Chapter 3 introduced the principles of statutory valuation for acquisition, including the statutory framework and process, with a focus on the concept of disregard of the scheme in compensation assessment and the distinction between concepts of value and worth. This provided an insight into concepts of fairness and equity in statutory valuation in the context of the disregard of compulsion.

Chapter 4 explored the market value bases of compensation in acquisition for which statutory valuation may be required, including market value of the interest and the related market value compensation concepts of severance and injurious affection. This provided a broad

Table 10.4 Hypothetical development method – example

1	*A B C*	*D*	*E*	*F*	*G*	*H*	*I*	*J*	*K*	*L*	*M*
2											
3		*Total*	*0*	*1*	*2*	*3*	*4*	*5*	*6*	*7*	*8*
4	**Net development value after sale costs**	$15,284,615									$15,284,615
5											
6	**Development costs**										
7	Land purchase	-$4,834,477	-$4,834,477								
8	Land purchase costs	-$241,724	-$241,724								
9	Site preparation	-$25,000	-$25,000								
10	Building costs	-$5,600,000			-$840,000	-$1,120,000	-$1,120,000	-$1,120,000	-$840,000	-$560,000	
11	External costs	-$75,000			-$37,500	-$37,500					
12	Professional fees	-$560,000			-$280,000	-$280,000					
13	Contingency allowance	-$560,000				-$140,000	-$140,000	-$140,000	-$140,000		
14	Letting agents fee	-$100,000									-$100,000
15	Letting legal fee	-$50,000									-$50,000
16	Developers profit	-$1,806,930									-$1,806,930
17	Net cash flow before finance		-$5,101,201	$0	-$1,157,500	-$1,577,500	-$1,260,000	-$1,260,000	-$980,000	-$560,000	$13,327,685
18											
19	**Finance**										
20	Opening balance			-$5,101,201	-$5,196,848	-$6,451,789	-$8,150,260	-$9,563,077	-$11,002,385	-$12,188,680	-$12,977,218
21	Interest at 1.875% per qtr			-$95,648	-$97,441	-$120,971	-$152,817	-$179,308	-$206,295	-$228,538	-$243,323
22	Closing balance		-$5,101,201	-$5,196,848	-$6,451,789	-$8,150,260	-$9,563,077	-$11,002,385	-$12,188,680	-$12,977,218	**$107,144**
23											

Source: Author

understanding of the heads of compensation for acquisition on a value basis that may arise in statutory valuation.

Chapter 5 considered non-market value forms of compensation, including special value, which was noted to rarely arise, payment for solace arising from loss of home which is heavily dependent on the exact wording of the facilitating statute and the concept of disturbance for costs and losses suffered as a result of the acquisition. This provided a broad understanding of the heads of compensation for acquisition that are not necessarily on a value basis but that may arise in statutory valuation.

Chapter 6 concluded the chapters considering statutory valuation for acquisition by examining compensation for other interests such as easements and mortgages, mineral rights and profits à prendre, strata title and substrata, businesses and the emerging statutory valuation areas of native title and of indigenous cultural loss. This provided an appreciation of the diversity of interests that may be subject to acquisition and so may require valuation for the assessment of compensation.

Statutory valuation for rating and taxing was considered in Chapters 7 and 8. Chapter 7 introduced the principles of land as the basis for taxation through land tax, stamp duty and rates, considering the general principles of the valuation of land for taxation using the New South Wales statutory framework as an example. This provided a general understanding of the often arcane and artificial basis of statutory valuation upon which government collects very significant amounts of taxation revenue.

Chapter 8 explored the practice of statutory valuation for rating and taxing through exemptions, concessions and allowances that have evolved over the last century, together with contemporary issues such as mass appraisal, automated valuation models and artificial intelligence, the impact of major disasters on valuation for rating and taxing and the litigation process, including the valuer as an expert witness. This provided an insight into the continued evolution of statutory valuation to address issues as they arise over time, highlighting the importance of a principles-based approach to statutory valuation.

Regulatory valuation was considered in chapters 9 and 10 in the context of the preparation of valuations for inclusion in financial reporting. Chapter 9 explored the principles of regulatory valuation with regard to the inter-related roles of international accounting standards, international financial reporting standards and international valuation standards, focusing on the primacy of fair value in international financial reporting standards. This provided an understanding of the fundamental conceptual differences and bases between statutory valuation and regulatory valuation.

With an emphasis on accounting standards, valuation standards and fair value for financial reporting purposes, regulatory valuation has a wider range of valuation approaches and methods available for adoption rather than the emphasis in statutory valuation on the market approach and the comparative method of valuation. This chapter explored the income approach to valuation through the capitalisation of income method and the discounted cash flow method, being widely used by real estate investment trusts, superannuation funds and other institutional investors for the valuation of property investments for inclusion in financial reporting.

Further, the cost approach to valuation was explored through the replacement and reproduction cost methods of valuation and through the static and dynamic hypothetical development methods of valuation. The usefulness to decision-makers of valuations based on the replacement and reproduction cost methods of valuation was considered, together with the inherent instability of the static and dynamic hypothetical development methods of valuation, being dependent on a wide range of interdependent input variables.

This chapter provided an appreciation of the wider range of valuation approaches and methods available to the valuer for regulatory valuation, reflecting that, unlike statutory valuation, there is no requirement for defence of the valuation in Court (negligence claims excepted).

With Chapters 1 and 2 introducing the concepts and principles of statutory valuation, Chapters 3, 4, 5 and 6 outlining statutory valuation for acquisition and Chapters 7 and 8 exploring statutory valuation for rating and taxing, Chapters 9 and 10 considered regulatory valuation and so conclude this book.

References

AASB (2015) *AASB 116 property, plant and equipment*, AASB, Melbourne
AASB (2019) *AASB 140 investment property*, AASB, Melbourne
API (2021a) *AVGP 302 valuations of real property, plant and equipment for use in Australian financial reports*, API, Deakin
API (2021b) *ANZVGP 111 valuation procedures – real property*, API, Deakin
Baum, A, Mackmin, D and Nunnington, N (2018) *The income approach to property valuation*, Routledge, Abingdon
Davidson, A (1978) *Parry's valuation and conversion tables*, Estates Gazette, London
IVSC (2024) *International valuation standards*, IVSC, London
Millington, A (2000) *An introduction to property valuation*, Routledge, Abingdon
Parker, D (2016) *International valuation standards – a guide to the valuation of real property assets*, Routledge, Abingdon
Parker, D (2022) Valuation principles, in D Parker (Ed) *Principles and practice of property valuation in Australia*, Routledge, Abingdon
Parker, D (2024) *Introduction to property valuation in Australia*, Routledge, Abingdon
Rawlinsons (2024) *Rawlinsons Australian construction handbook 2024*, Rawlinsons, Perth
RICS (2024) *RICS valuation – global standards (red book)*, RICS, London
Rost, RO and Collins, HG (1984) *Land valuation and compensation in Australia*, 3rd Edition, Australian Institute of Valuers and Land Administrators, Deakin
Stewart, R (2022) Valuation for financial reporting purposes, in D Parker (Ed) *Principles and practice of property valuation in Australia*, Routledge, Abingdon
Wyatt, P (2023) *Property valuation*, Wiley Blackwell, Chichester

Index

Note: numbers in **bold** indicate a table